AF863199

Advances in Industrial Control

Series Editors

Michael J. Grimble, Industrial Control Centre, University of Strathclyde, Glasgow, UK

Antonella Ferrara, Department of Electrical, Computer and Biomedical Engineering, University of Pavia, Pavia, Italy

Editorial Board

Graham Goodwin, School of Electrical Engineering and Computing, University of Newcastle, Callaghan, NSW, Australia

Thomas J. Harris, Department of Chemical Engineering, Queen's University, Kingston, ON, Canada

Tong Heng Lee, Department of Electrical and Computer Engineering, National University of Singapore, Singapore, Singapore

Om P. Malik, Schulich School of Engineering, University of Calgary, Calgary, AB, Canada

Kim-Fung Man, City University Hong Kong, Kowloon, Hong Kong

Gustaf Olsson, Department of Industrial Electrical Engineering and Automation, Lund Institute of Technology, Lund, Sweden

Asok Ray, Department of Mechanical Engineering, Pennsylvania State University, University Park, PA, USA

Sebastian Engell, Lehrstuhl für Systemdynamik und Prozessführung, Technische Universität Dortmund, Dortmund, Germany

Ikuo Yamamoto, Graduate School of Engineering, University of Nagasaki, Nagasaki, Japan

Advances in Industrial Control is a series of monographs and contributed titles focusing on the applications of advanced and novel control methods within applied settings. This series has worldwide distribution to engineers, researchers and libraries.

The series promotes the exchange of information between academia and industry, to which end the books all demonstrate some theoretical aspect of an advanced or new control method and show how it can be applied either in a pilot plant or in some real industrial situation. The books are distinguished by the combination of the type of theory used and the type of application exemplified. Note that "industrial" here has a very broad interpretation; it applies not merely to the processes employed in industrial plants but to systems such as avionics and automotive brakes and drivetrain. This series complements the theoretical and more mathematical approach of Communications and Control Engineering.

Indexed by SCOPUS and Engineering Index.

Proposals for this series, composed of a proposal form downloaded from this page, a draft Contents, at least two sample chapters and an author cv (with a synopsis of the whole project, if possible) can be submitted to either of the:

Series Editors

Professor **Michael J. Grimble**
Department of Electronic and Electrical Engineering, Royal College Building, 204 George Street, Glasgow G1 1XW, UK
e-mail: m.j.grimble@strath.ac.uk

Professor **Antonella Ferrara**
Department of Electrical, Computer and Biomedical Engineering, University of Pavia, Via Ferrata 1, 27100 Pavia, Italy
e-mail: antonella.ferrara@unipv.it

or the

In-house Editor

Mr. **Oliver Jackson**
Springer London, 4 Crinan Street, London, N1 9XW, UK
e-mail: oliver.jackson@springer.com

Proposals are peer-reviewed.

Publishing Ethics

Researchers should conduct their research from research proposal to publication in line with best practices and codes of conduct of relevant professional bodies and/or national and international regulatory bodies. For more details on individual ethics matters please see: https://www.springer.com/gp/authors-editors/journal-author/journal-author-helpdesk/publishing-ethics/14214

More information about this series at http://www.springer.com/series/1412

Mirko Mazzoleni · Gianpietro Di Rito · Fabio Previdi

Electro-Mechanical Actuators for the More Electric Aircraft

Springer

Mirko Mazzoleni
Department of Management, Information and Production Engineering
University of Bergamo
Bergamo, Italy

Gianpietro Di Rito
Department of Civil and Industrial Engineering
University of Pisa
Pisa, Italy

Fabio Previdi
Department of Management, Information and Production Engineering
University of Bergamo
Bergamo, Italy

ISSN 1430-9491 ISSN 2193-1577 (electronic)
Advances in Industrial Control
ISBN 978-3-030-61798-1 ISBN 978-3-030-61799-8 (eBook)
https://doi.org/10.1007/978-3-030-61799-8

MATLAB and Simulink are registered trademarks of The MathWorks, Inc. See https://www.mathworks.com/trademarks for a list of additional trademarks.

Mathematics Subject Classification (2010): 93E10, 93E12, 60G35, 62H25, 68T10, 62H30, 93A30

© Springer Nature Switzerland AG 2021
This work is subject to copyright. All rights are reserved by the Publisher, whether the whole or part of the material is concerned, specifically the rights of translation, reprinting, reuse of illustrations, recitation, broadcasting, reproduction on microfilms or in any other physical way, and transmission or information storage and retrieval, electronic adaptation, computer software, or by similar or dissimilar methodology now known or hereafter developed.
The use of general descriptive names, registered names, trademarks, service marks, etc. in this publication does not imply, even in the absence of a specific statement, that such names are exempt from the relevant protective laws and regulations and therefore free for general use.
The publisher, the authors and the editors are safe to assume that the advice and information in this book are believed to be true and accurate at the date of publication. Neither the publisher nor the authors or the editors give a warranty, expressed or implied, with respect to the material contained herein or for any errors or omissions that may have been made. The publisher remains neutral with regard to jurisdictional claims in published maps and institutional affiliations.

This Springer imprint is published by the registered company Springer Nature Switzerland AG
The registered company address is: Gewerbestrasse 11, 6330 Cham, Switzerland

To Gianmarco, who guides me everyday
to discover a man I'd never dared to be.

Gianpietro Di Rito

Series Editor's Foreword

Control engineering is viewed rather differently by researchers and those that must implement and maintain control systems. Researchers develop general algorithms with a strong mathematical basis, whilst practitioners have more local concerns over the capabilities of equipment, quality of control and plant downtime. The series *Advances in Industrial Control* attempts to bridge this divide and hopes to encourage the adoption of advanced control techniques for applications where they can boost safety, quality of control and profitability.

The rapid development of new control theory and technology has an impact on all areas of engineering. This monograph series has a focus on applications, since they are the challenges of an industry that stimulate the development of new control paradigms. The questions of "control design" explored in the series have often been relegated to the second division of controls research. A greater focus on applications is desirable if the different aspects of the "control design" problem are to be explored with the same dedication that "control synthesis" problems have received in the past. It is hoped that the series will cover the substantial benefits that advanced control can provide whilst tempering enthusiasm by addressing the challenges that can arise.

This monograph covers *Electro-Mechanical Actuators for the More Electric Aircraft* with reference to the application of condition monitoring and fault diagnosis. It is timely since in the aftermath of the coronavirus pandemic the world will be looking for new developments that lead to a safer and healthier world where the environment has a higher priority. Aircraft electrification provides many opportunities to optimize energy, improve efficiency, reduce weight and costs, and provide greater flexibility in designs at the same time as improving safety. The benefits of Fly-By-Wire systems are now of course well-known and are accepted.

The authors provide a wide-ranging introduction to the subject and to the current state of developments in Chap. 1. The material is very timely particularly on environmental and societal issues. The impact and importance of electrically powered actuators are also covered in this chapter, as are the important topics of power and control electronics. It is refreshing that the text also covers the various aircraft sub-systems; such coverage is useful for engineers working on real aircraft systems. Many of us live in the world of simulation and need greater exposure to the real limits of equipment and devices.

Chapter 2 turns to the reliability and safety of airborne electro-mechanical actuators. However, much of the terminology and many of the ideas apply to many applications, so the text has a wider reach than simply aerospace systems. Fault tolerance is of course of greater importance in this industry than most. Engineers concerned with aircraft safety systems should find the material very valuable since it is not so accessible with a control engineering focus elsewhere.

Fault diagnosis, fault estimation, fault identification and condition monitoring are important tools that will be employed more extensively in future systems. Chapter 3 introduces the basic concepts and goes on to describe various approaches to implementation. The use of fault accommodation, analytical redundancy and reconfiguration has been discussed many times from a theoretical viewpoint, but the text provides some hope that these methods will be employed in real aircraft systems. The use of model-based methods involving the ubiquitous Kalman filtering or parameter identification-based schemes is described. Some of the topics relate to various areas in signal processing and artificial intelligence.

Chapter 4 considers fault diagnosis problems for airborne electro-mechanical actuators, a topic that is important to the electrification of aircraft systems. The problems are first described from a rather practical viewpoint, including descriptions of electrical equipment. The modelling and simulation of problems are described, and a model-based approach is explained in some detail. The alternative signal-based or knowledge-based approaches are also discussed.

The text is a welcome addition to the series and is unusual since an engineer in the aerospace industry should find the material as accessible as an academic or research scientist. The argument for more electric aircraft systems speaks for itself, and the need to make aircraft more environmentally friendly must be an aim of all manufacturers and airlines.

Glasgow, UK
June 2020

Michael J. Grimble

Preface

Engineering systems are subject to faults. The early detection of these abnormal occurrences is of paramount importance from different points of view, which range from assuring product quality in manufacturing processes to safety concerns in situations where a damage to machineries and humans could be possible. Modern technology is characterized by the interconnection of many automated components, which interact in complex ways: the detection and accommodation of a faulty component can avoid the propagation of the fault to the whole system.

The increase in system complexity involved also the aerospace case, due to major requirements in range, speed, and control functions needed for modern aircraft. This implied a significant increase in maintenance costs for hydraulic and pneumatic systems. Electrically-powered systems do not suffer from many of the inherent shortcomings of hydraulic, pneumatic, and mechanical ones: they are relatively flexible and light, more environmentally sustainable, and have higher efficiency. Thanks to industrial and research investments pursuing the More Electric Aircraft (MEA) initiative, the technological readiness of electric systems is nowadays concrete.

A key factor for achieving the MEA objectives is the use of electrically-powered actuation systems. Electro-Mechanical Actuators (EMAs) remove completely the need for hydraulic power, thus reducing the environmental imprint, the weight and space volumes needed for their installation. A critical issue to be addressed in the development of aircraft EMAs is the management of the fail-safe mode of the system. In hydraulic actuation, these protection functions were effectively and efficiently accomplished via hydraulic components, while in EMAs they must be implemented by mechanical, electromagnetic, or electric devices. Fault-tolerant EMA systems were developed for this reason, where the robustness to faults is implemented in different levels of the actuator (i.e. power electronics, motor, and transmission).

Electro-Mechanical Actuators are usually paired with an Electronic Control Unit (ECU) that takes care of the EMA control. A specific portion of the ECU can, therefore, be devoted to monitoring tasks. The fault diagnosis algorithm has the duty of detecting abnormalities in the actuator operations. This analytical redundancy capability, added to the hardware redundancy of fault-tolerant architectures, further enhances the ability of the actuation system to cope promptly with faults.

Scope. The aim of this book is to present algorithmic approaches to the fault diagnosis and condition monitoring of airborne EMAs. The first three chapters set the stage for the remaining content of the monograph, by introducing the MEA concept and related issues, the Reliability, Availability, Maintainability and Safety (RAMS) discipline, and diagnosis approaches.

The book is written with the idea of giving a practical approach to fault diagnosis and monitoring or flight EMAs. The fourth chapter presents validated diagnosis methods that make use of different rationales: model-based, signal-based, and knowledge-based approaches. The last chapter contains notes for practitioners, learned from the experience of the authors, in developing diagnostic solutions in the aerospace sector.

The book can be of interest for researchers in automatic control, aerospace and mechanical engineering dealing with fault diagnosis problems, but also for the practitioner working in industrial sectors.

Outline of the book. The book is structured as follows:

- Chapter 1 introduces the more electric aircraft initiative, reviewing the trends in the development of electrically-powered systems for aerospace applications. The state of the art of electro-mechanical actuation systems in aircraft is presented.
- Chapter 2 presents the concepts of reliability, availability, maintainability, and safety analysis for aircraft applications. A practical example concerning an electro-mechanical actuation system for morphing flaps is given.
- Chapter 3 describes the terminology and the main approaches for fault diagnosis and condition monitoring. A specific section is devoted to the application of these algorithms to electro-mechanical actuators.
- Chapter 4 shows various applications of fault diagnosis and condition monitoring to aerospace electro-mechanical actuators. Different strategies are presented, following the treatment done in Chap. 3.
- Chapter 5 is devoted to concluding remarks, lessons learned, and suggestions for future works.

Stay healthy. Fault diagnosis methods rely on the generation of fault indicators. When an engineering system operates in its normal behavior, those indicators lie in a nominal range of values. When a fault occurs, it is desirable that the indicators deviate from their nominal value.

No system is complex as the human body, and it provides several symptoms that something is deviating from the nominal healthy conditions (e.g. fever, cough, …). During the creation of this book, the humankind was threatened by the SARS-CoV-2 pandemic. Italy, and in particular the city of Bergamo, was one of the most hit states in Europe: we personally know at least one person that was carried away by the virus. It is in moments like this that one may wonder: what is of primary importance? What is the purpose of technological progress if an invisible microscopic entity can break opulences and societies? There is no easy answer: surely, humankind has always been able to rise from its ashes.

Maybe, to alleviate the grief of having lost a loved one, every time we see an airplane cutting through the sky with flames from its propulsion engines like if it was a phoenix, we can pretend that this is the soul of our beloved ones, watching over us for the times to come.

Bergamo, Italy — Mirko Mazzoleni
Pisa, Italy — Gianpietro Di Rito
Bergamo, Italy — Fabio Previdi
August 2020

Acknowledgements The authors express their gratitude to the European Union for financial support for the HOLMES and REPRISE projects presented in the book. These permitted us to collaborate with top-level industries in the field and learning from other people. We thank the UmbraGroup, Piaggio Aerospace, Zettlex, Mecaer, Liebherr Aerospace, and Leonardo Velivoli companies for giving us the possibility to work on important topics.

Finally, we would like to express our gratitude to Oliver Jackson from Springer Nature and the Series Editor for their valuable support.

Contents

Abbreviations

A/C	Aircraft
AAEP	Active/Active Equal Power
AAPR	Active/Active Pure Rotation
AAPT	Active/Active Pure Translation
AC	Alternate Current
ACU	Actuator Control Unit
AEA	All Electric Aircraft
AEP	All Electric Propulsion
AF	Angle Feedback
ANC	Adaptive Noise Cancellation
APU	Auxiliary Power Unit
ASB	Active/Stand-By
AUC	Airborne Uninhabited Cargo
BIT	Built-in Test
BLACM	Brushless Alternate Current Machine (or Motor)
BLDCM	Brushless Direct Current Machine (or Motor)
BPC	Battery Power Coverage
BSE	Battery Specific Energy
CA	Criticality Analysis
CAF	Consolidated Angle Feedback
CAGR	Compound Annual Growth Rate
CAN	Controller Area Network
CBIT	Continuous Built-in Test
CCA	Common Cause Analysis
CK	Correlated Kurtosis
CM	Condition Monitoring
CON	CONtrol electronic unit
CPU	Central Processing Unit
CS2	Clean Sky 2
CVA	Canonical Variate Analysis

DAL	Development Assurance Level
DC	Direct Current
DFT	Discrete Fourier Transform
DQZ	Direct-Quadrature-Zero
DTW	Dynamic Time Warping
Eq/F	Equipment/Furnishing
EAF	Estimated Angle Feedback
EBHA	Electro-Backup-Hydrostatic Actuator
EC	European Commission
ECS	Environmental Control System
ECU	Electronic Control Unit
EGTS	Electric Green Taxiing System
EHA	Electro-Hydrostatic Actuator
EHA-FD	Fixed-Displacement EHA
EHA-VD	Variable-Displacement EHA
EKF	Extended Kalman Filter
EMA	Electro-Mechanical Actuator
EMS	Energy Management System
EPGDS	Electrical Power Generation and Distribution System
ESM	Electrically-excited Synchronous Machine
EU	European Union
EVTOL	Electric Vertical Take-Off and Landing
FBW	Fly-By-Wire
FCC	Flight Control Computer
FCS	Flight Control System
FD	Fault Detection
FDI	Fault Detection and Isolation
FDIA	Fault Detection, Isolation and Analysis
FDL	Fault Detection Logic
FFT	Fast Fourier Transform
FHA	Functional Hazard Assessment
FL	Fuzzy Logic
FM	Failure Mode
FMEA	Failure Mode and Effect Analysis
FMECA	Failure Mode Effects and Criticality Analysis
FOC	Field Oriented Control
FP	Fault Prognosis
FP6	6th Framework Program (EC funding program)
FP7	7th Framework Program (EC funding program)
FTA	Fault Tree Analysis
FTK	Freight Tonne Kilometre index
GPIO	General-Purpose Input/Output
H2020	Horizon 2020 (EC funding program)
H/C	HeliCopter
HHT	Hilbert–Huang Transform

HMI	Human–Machine Interface
HOLMES	Health On Line Monitoring for Electro-MEchanical actuator Safety
HW	HardWare
IAP	Integrated Actuator Package
IBIT	Initialising Built-in Test
ICA	Independent Component Analysis
ILM	Inner Loop Monitor
IM	Induction Machine (or Motor)
IPS	Ice Protection System
JDL	Jamming Detection Logic
JTI	Joint Technology Initiative
JTU	Joint Technology Undertaking
KF	Kalman Filter
LF	Linear Feedback
LRU	Line Replacement Unit
LTI	Linear Time Invariant
LVDT	Linear Variable Differential Transformer
MALE	Medium Altitude Long Endurance
MBIT	Maintenance Built-in Test
MCC	Most Critical Conditions
MCSA	Motor Current Signature Analysis
MEA	More Electric Aircraft
MED	Minimum Entropy Deconvolution
MEP	More Electrical Propulsion
MFS	Morphing Flap System
MLG	Main Landing Gear
MON	MONitor electronic unit
MPE	Motor Power Electronics
MTBF	Mean-Time Between Failures
MTOW	Maximum Take-Off Weight
MOET	More Open Electrical Technologies
MOSFET	Metal–Oxide–Semiconductor Field-Effect Transistor
NLG	Nose Landing Gear
OLM	Outer Loop Monitor
PBIT	Power-up Built-in Test
PBW	Power-By-Wire
PCA	Principal Component Analysis
PCM	Prognostic Condition Monitoring
PD	Partial Discharge
PDF	Probability Density Function
PEU	Power Electronic Unit
PF	Particle Filter
PLS	Partial Least Squares
PMSM	Permanent Magnet Synchronous Machine (or Motor)
PSSA	Preliminary System Safety Assessment

PSU	Power Supply Unit
PTM	Position-Tracking Monitor
PWM	Pulse-Width Modulation
QTA	Qualitative Trend Analysis
R&D	Research and Development
RAMS	Reliability, Availability, Maintainability and Safety
RBD	Reliability Block Diagram
REPRISE	Reliable Electro-mechanical actuator for PRImary SurfacE with health monitoring
RPK	Revenue Passenger Kilometre index
RTCA	Radio Technical Commission for Aeronautics
RUL	Remaining Useful Life
RVDT	Rotary Variable Differential Transformer
SBA	Stand-By/Active
SESAR	Single European Sky Air traffic management Research
SHA	Servo-Hydraulic Actuator
SIFT	Scale-Invariant Feature Transform
SIM	Subspace Identification Methods
SK	Spectral Kurtosis
SKR	Stable Kernel Representation
SMC	Sequential Monte Carlo
SPI	Serial Peripheral Interface
SPM	Statistical Process Monitoring
SRM	Switched Reluctance Machine (or Motor)
SSA	System Safety Assessment
STFT	Short-Time Fourier Transform
SVPWM	Space-Vector Pulse-Width Modulation
SW	SoftWare
TCP	Transmission Control Protocol
TMR	Triple Modular Redundancy
TRL	Technology Readiness Level
TRU	Transformer Rectifier Unit
TSA	Time Synchronous Averaging
TVC	Thrust Vectoring Control
UAS	Unmanned Aerial System
UAV	Unmanned Aerial Vehicle
UIO	Unknown Input Observer
UKF	Unscented Kalman Filter
VAF	Voted Angle Feedback
VTOL	Vertical Take-Off and Landing
WMD	Wigner–Ville Distribution
WT	Wavelet Transform

Chapter 1
Introduction

Outline of the chapter. The first chapter of the book has the objective to present the research and development framework aiming at the so-called More Electric Aircraft (MEA) concept, with a specific focus on Electro-Mechanical Actuators (EMAs). Section 1.1 points out the technological, environmental, societal, and market impacts of the MEA concept. Section 1.2 describes the most relevant developments of EMA technologies, from hybrid electro-hydraulic solutions to the current EMA state of the art. Section 1.3 reviews the application of EMAs to specific functionalities of the aircraft, ranging from the most conventional to more innovative ones. Finally, Sect. 1.4 summarizes the content of the chapter.

1.1 Electrification of Onboard Power Systems: The "More Electric Aircraft" Concept

Aviation has fundamentally transformed society over the past 40 years. The economic and social benefits gained by the efficient and fast transportation of people and goods led to an overwhelming growth of air traffic over the past 20 years, and this trend was expected to continue in the future, particularly for the growing markets of the Far East. In 2019, considering only commercial airlines, according to IATA [3], the global fleet included 29697 airplanes, with about 4.5 million available seats and 2.9 million jobs. In the same year, passenger trips on U.S.A. airlines were 925 million, the highest value ever, with a record occupancy rate of 84.6%. Unfortunately, the apparently relentless growth of the market faced a sudden stop due to the Sars-CoV-2 pandemic [9] that had catastrophic effects on the world economy and even more devastating impact on aerospace and defense industry that in 2020 was "facing probably the gravest crisis in its history", according to Guillaume Faury, Chief Executive Officer of the aircraft maker Airbus [34].

So, it must be a paramount concern to continuously support important and strategic initiatives for the innovation of the aviation industry, which have been worldwide launched in the past decades with the common target of optimizing the

© Springer Nature Switzerland AG 2021

M. Mazzoleni et al., *Electro-Mechanical Actuators for the More Electric Aircraft*, Advances in Industrial Control, https://doi.org/10.1007/978-3-030-61799-8_1

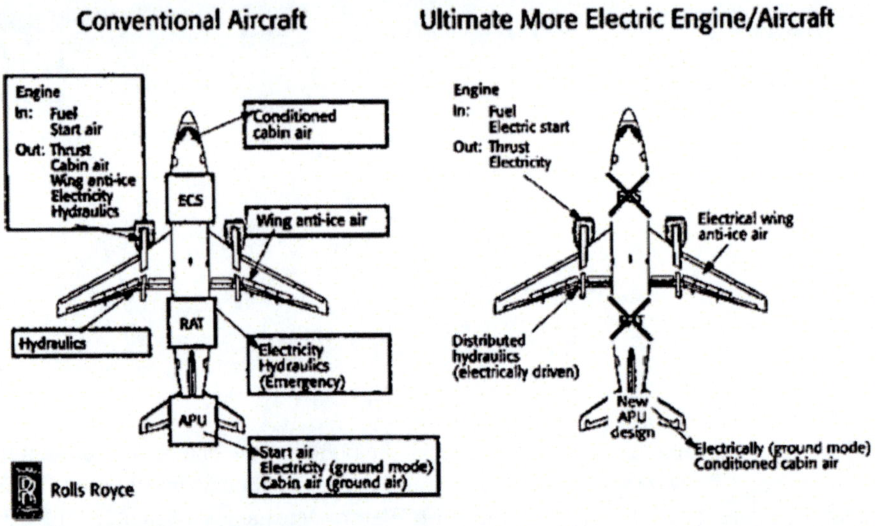

Fig. 1.1 Aircraft power systems in conventional and more electric concepts. Republished with permission of Institution of Engineering and Technology (IET), from [35]: All electric aircraft, Howse, M., 17(4) © 2003; permission conveyed through Copyright Clearance Center, Inc

performances, the power efficiency, the maintainability, the reliability/safety, and the eco-compatibility of aircraft. In this context, a major interest has been focused on a design philosophy named *aircraft electrification*, which manifests in two basic concepts:

- *More Electric Aircraft* (MEA), see Fig. 1.1, pursuing the long-term target of *All Electric Aircraft* (AEA), which entails the gradual replacement of onboard systems based on mechanical, hydraulic, or pneumatic power sources with electrically powered systems [29, 35, 56].
- *More Electrical Propulsion* (MEP), pursuing the long-term target of *All Electric Propulsion* (AEP), which can potentially imply a kind of revolution in the whole aircraft design approach, and could transform large segments of the aerospace industry, by affecting not only propulsion but also aircraft systems [51, 52].

With particular reference to EU, it is worth mentioning the collaborative research initiatives within the EC-funded programs FP6, FP7, and Horizon 2020, such as the MOET program [6, 7], the SESAR Joint Undertaking [8], the Clean Sky Joint Technology Initiative [1], as well as national programs and private company programs [10, 46], all aiming to improve the Technology Readiness Level (TRL) of electrically powered systems.

Table 1.1 Engine power output in kW for the A330 aircraft. Source: Roland Berger, https://www.rolandberger.com/publications/publication_pdf/roland_berger_aircraft_electrical_propulsion.pdf

Electrical generator	High-pressure bleed air (pneumatic)	Hydraulic pump	Fuel and oil pump (mechanical)	Thrust power	Total non-thrust power
200	1200	240	100	40000 ca.	1700 ca.

1.1.1 Technological Issues

In the conventional design of aircraft systems, non-propulsive functions such as actuation, de-icing, and air-conditioning utilize mechanical, hydraulic, and pneumatic power sources, extracted by the aircraft engines via a variety of mechanisms (hydraulic and electric power is derived from geared mechanical transmissions, while pneumatic power is obtained by air bleeding of the engine compressor). Almost all the engine power is used for thrust, while the non-propulsive functions typically absorb 5% of the total power [52], see Table 1.1.

The increases in range, speed, and control functions needed for modern aircraft have clearly led to the increase of complexity of onboard systems. This implied a significant increase of maintenance costs for hydraulic and pneumatic systems, requiring to check long, complex, and heavy pipes and ducts running throughout the aircraft. In addition, pneumatic systems have low-power efficiency, and hydraulic systems require heavy heat exchangers to maintain the fluid at an adequate operating temperature. Electrically powered systems do not suffer from many of the inherent shortcomings of hydraulic, pneumatic, and mechanical ones: they are relatively flexible and light, and have higher efficiency.

A key milestone in the trend to the MEA was the introduction of a Fly-By-Wire (FBW) flight control system in the Airbus A320 in the late 1980s, followed by the Boeing 777 in 1994. The FBW technology significantly reduced weight and provided additional space for other aircraft components by enabling the electrical transmission of commands from the cockpit to the flight controls. The next big step came with the development of the A380 [58], and the implementation of an electrically actuated thrust reverser, along with use of electrically powered actuators for some wing and tail flight controls, see Figs. 1.2 and 1.3. Finally, the Boeing 787 was the first large transport aircraft to have an electrically powered environmental control system, and to employ electrically actuated brakes and electrical de-icing. In the military sector, the JSF F-35 employs, thanks to the use of high-voltage DC distribution system, a fully electrically powered flight control actuation system [51], Fig. 1.4. Together with the elimination of the problems related to hydraulic and pneumatic systems, the more electric solution enables an easier system integration, and it implies a strong increase of flexibility in terms of size, shape, and location of aerodynamic control surfaces.

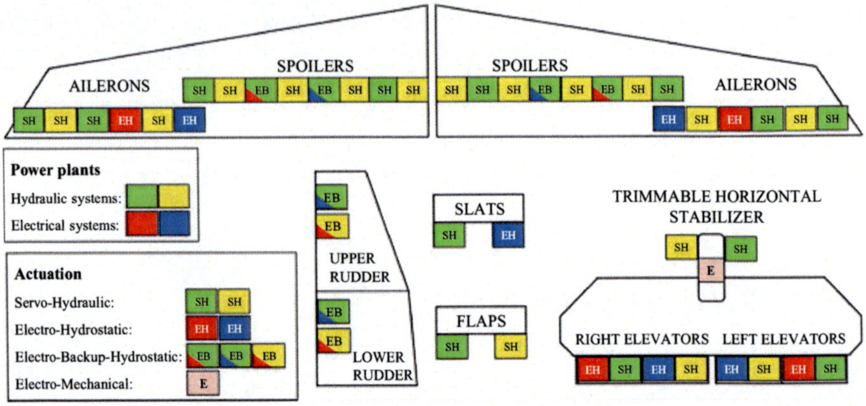

Fig. 1.2 A380 actuation system power distribution. (Green-Yellow) The two hydraulic systems; (Red-Blue) the two electric systems

Fig. 1.3 A380 flight control surfaces. Adapted from [45]—originally published open access and licensed under CC-BY 4.0. https://ieeexplore.ieee.org/stamp/stamp.jsp?tp=&arnumber=8878102

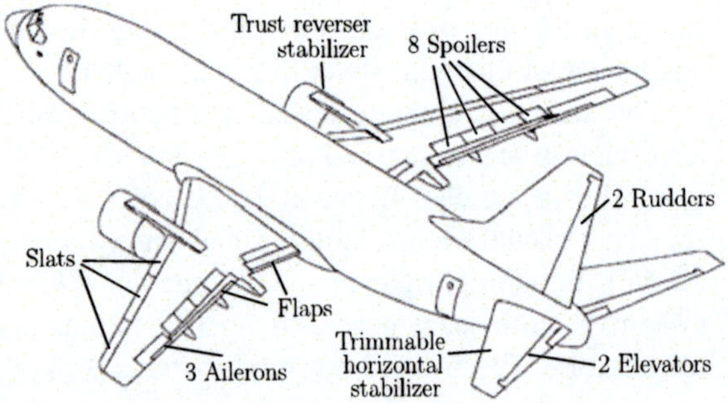

The main potential results provided by the MEA have been recognized as

- Reduction of weight at aircraft-level,
- Energy optimization, through
  - increased efficiency of power systems;
  - "power-on-demand" capability;
  - reduction of fuel consumption;
- Increase of environmental compatibility, through
  - waste reduction (thanks to the elimination of hydraulic fluids);
  - reduction of CO_2-emissions;
- Increase of survivability, reliability, maintainability and safety, through:
  - simplified integration;
  - reduced inflame danger (thanks to the elimination of hydraulic fluids);
- Reduction of operative and maintenance costs.

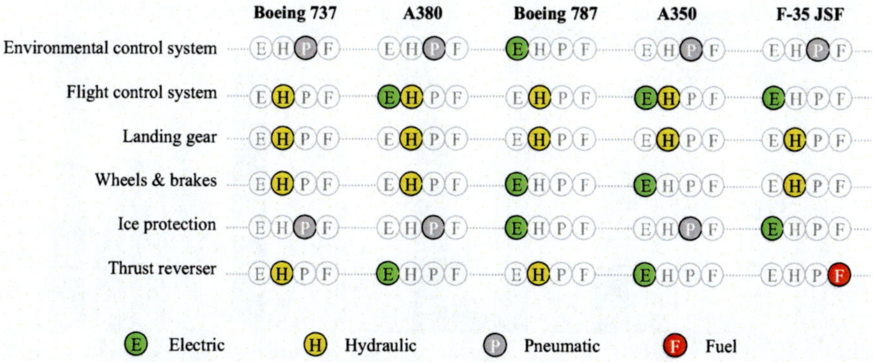

Fig. 1.4 Technological solutions for onboard power systems in some reference aircraft. Adapted with permission from Roland Berger. *Source* Airbus, Boeing, Lockheed Martin, Roland Berger, https://www.rolandberger.com/publications/publication_pdf/roland_berger_aircraft_electrical_propulsion.pdf

Many research investigations over the past 40 years have demonstrated the validity of the MEA concept, and recent outcomes confirm that the use of all electric technologies for long-range civil aircraft is expected to obtain up to 10% reduction in empty weight and 9% reduction in fuel consumption [21, 50].

Nevertheless, the TRL of electrically powered systems has been poor up to the late 1990s, and conventional systems were preferred to the more electric ones. Thanks to industrial and research investments, the TRL of electric systems has been more and more enhanced, and nowadays the technological, economic, and environmental impacts of the MEA can be concrete [40]. The Boeing 787 Dreamliner is probably the best example of MEA initiative: its electrical loads absorb almost 1000 kVA compared to the 300 kVA of a more conventional Airbus A320 [50].

The MEA concept clearly implies the increase of electric generation capacity on aircraft, and this point must be addressed with attention when evaluating the technological impacts of on-board systems electrification [52], Fig. 1.5. Actually, to avoid oversized electrical generators, the Electrical Power Generation and Distribution System (EPGDS) of an MEA must include an energy management control logic capable of monitoring and managing the electrical power requests, in order to minimize overloads and/or possible lack of energy for safety-critical functions (e.g., flight controls).

Relevant R&D activities have been carried out in the framework of CleanSky JTU programs (*CleanSky* within FP7, and successively *CleanSky2* within H2020) with reference to the MEA topics. A particular focus was made on energy management concerns for regional aircraft applications (Fig. 1.6). By using a shared simulation platform (including detailed dynamic models of onboard systems developed with different approaches and software languages, e.g., AMESim, Modelica-Dymola, MATLAB-Simulink), the power flows of the MEA systems were characterized in steady-state and transient conditions, simulating a number of flight maneuvers in different mission phases [55].

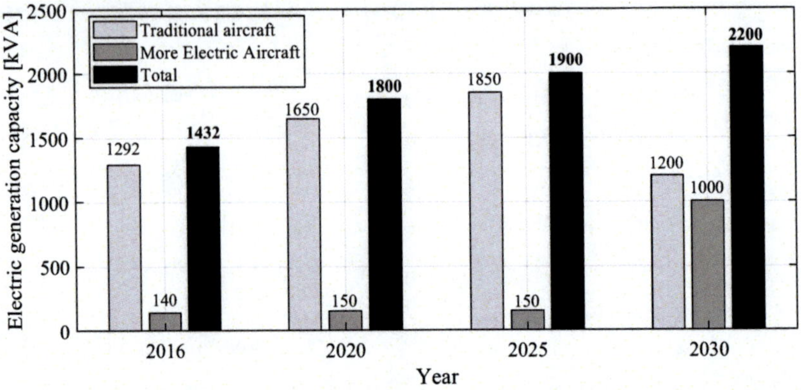

Fig. 1.5 Increase of electric power generation for large transport aircraft. Adapted with permission from Roland Berger. *Source* Teal, Roland Berger, https://www.rolandberger.com/publications/publication_pdf/roland_berger_aircraft_electrical_propulsion.pdf

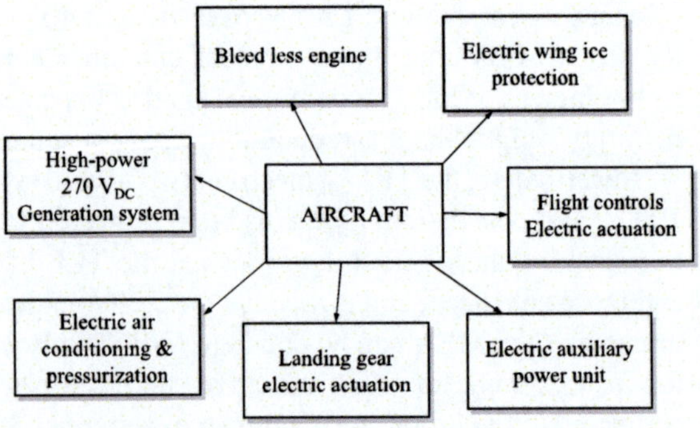

Fig. 1.6 Clean Sky "Green Regional Aircraft" power systems

Some results of these studies are reported in Fig. 1.7: the upper plots show the total absorbed power and the power requests from the Ice Protection System (IPS), Equipment/Furnishing (Eq/F), Internal Lights, Entertainment, and Environmental Control System (ECS), while the lower plots report the electrical voltages supplying the systems. It can be noted that when the total power reaches an activation threshold (105 kW), the Energy Management System (EMS) fades the voltages of noncritical loads and commands the ECS to reduce its power absorption (to about 50 kW). When the first power reduction of the ECS is achieved, the total absorption is still over the threshold, so the ECS power is again reduced and the noncritical loads are re-energized. Afterward, the EMS increases again the power for the ECS until the activation threshold is reached once more. This dynamics leads to low-frequency limit-cycle oscillations of the total power absorption, which in any case have minor impacts on the aircraft cabin parameters, Fig. 1.8.

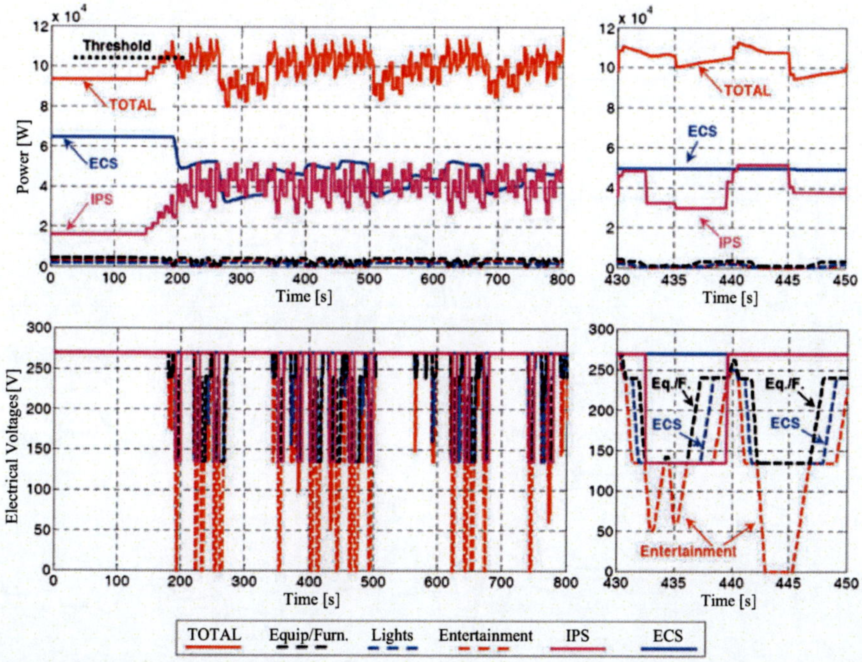

Fig. 1.7 Clean Sky "Green Regional Aircraft" simulation: systems' power. Schettini, F., Denti, E., Di Rito, G., Development of a simulation platform of all-electric aircraft on-board systems for energy management studies, The Aeronautical Journal 121, 1239, page 710–719 (2018), reproduced with permission

If the MEA concept can be viewed as a today reality and the MEP as a challenging but affordable target, the transition to AEA/AEP applications is still far to be concrete, especially for the large transport aircraft category. As highlighted by the data in Table 1.2, the limitation is essentially due to the specific power of batteries: by assuming a Battery-Specific Energy (BSE) of 300 Wh/kg [18] and that the fossil fuel mass on all aircraft contributes to the 30% of the total mass M_{AC}, we have that the Battery Power Coverage (BPC) is roughly given by (1.1):

$$\mathrm{BPC} = \mathrm{BSE} \cdot \frac{k_f M_{\mathrm{AC}}}{P_d D_m}, \tag{1.1}$$

where k_f is 0.3, D_m is the mission duration in hours and P_d is the total power demand in kW, including propulsion. At the current state of battery technology, the results point out that the all electric configuration can be considered feasible only for the urban mobility category. Figure 1.9 compares different batteries technologies.

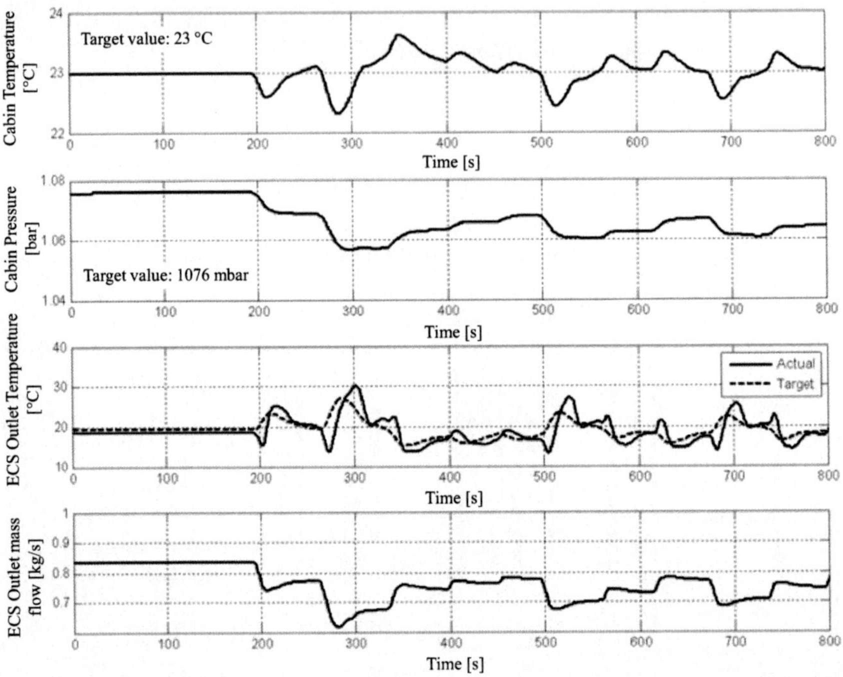

Fig. 1.8 Clean Sky "Green Regional Aircraft" simulation: cabin temperature. Schettini, F., Denti, E., Di Rito, G., Development of a simulation platform of all-electric aircraft on-board systems for energy management studies, The Aeronautical Journal 121, 1239, page 710–719 (2018), reproduced with permission

Table 1.2 Battery power coverage for different categories of aircraft, by assuming BSE = 300 Wh/kg

	Urban mobility	Regional	Medium-range	Long-range
Passengers	from 1 to 5	from 20 to 120	$\approx$180	$\approx$350
Range [nm]	<50	<500	<1000	>1000
Power demand (P_d) [kW]	from 250 to 500	from 1500 to 2500	from 20000 to 100000	from 200000 to 350000
A/C mass (M_{AC}) [kg]	from 2500 to 5000	from 10000 to 25000	from 50000 to 250000	from 250000 to 400000
Mission duration (D_m) [h]	1	3	5	10
Fuel mass factor (k_f)	0.3	0.3	0.3	0.3
Battery power coverage (BPC)	$\approx$0.90	$\approx$0.2	$\approx$0.05	$\approx$0.01

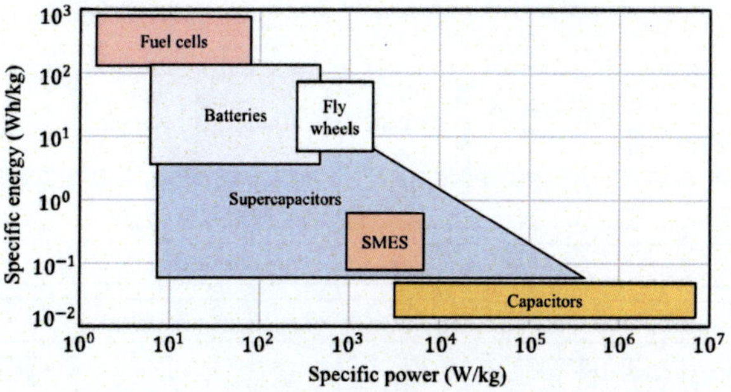

Fig. 1.9 Comparison of different batteries technologies

Table 1.3 Aviation's decades mean percentage impact on global CO_2-emissions. Source: Atmospheric Environment, IPCC, Roland Berger, https://www.rolandberger.com/publications/publication_pdf/roland_berger_aircraft_electrical_propulsion_2.pdf

1940–1950	1950–1960	1960–1970	1970–1980	1980–1990	1990–1900	2000–2010	2010–2020
0.88%	1.25%	1.77%	2.09%	2.31%	2.47%	2.57%	2.56%

1.1.2 Environmental and Societal Issues

The MEA/MEP concepts are thus strictly related to aviation sustainability, which has become a key factor for the innovation and the market orientation of the aerospace industry. Actually, though the environmental performance of aviation has improved dramatically since the 1960s, with the EU air traffic expected to increase from 9.5 million flights in 2012 to nearly 14.4 million in 2035, the challenge is meeting this growth in demand while minimizing its environmental impact. Table 1.3 shows the impact of aviation over the past 80 years on the global CO_2-emissions, where it can be noted that after a steady increase up to 1990s, the CO_2-emissions are essentially constant in the past 20 years, as a result of the industrial efforts to facilitate aviation sustainability [51, 54].

R&D activities on MEA/MEP topics aim to also invert the trend. As a relevant example, one of the main target expected from the SESAR JTU initiative is to enable a 10% reduction of the CO_2-emissions before 2035, i.e., a reduction from 250 to 500 kg of fuel burn per flight, Fig. 1.10.

Even more specific targets are given within the CleanSky JTU program, where the objectives are differentiated with respect to the vehicle category. As depicted in Fig. 1.11, the reduction of CO_2-emissions is expected to be relevant for long-range aircraft (−19%), and up to prodigious for short to medium-range aircraft and rotorcraft (−41% and −59% CO_2-emissions, respectively). The impacts in terms of noise abatement are also expected very good (more than −5 dB for all transport categories), with a special advance for regional turboprop aircraft (−15.7 dB).

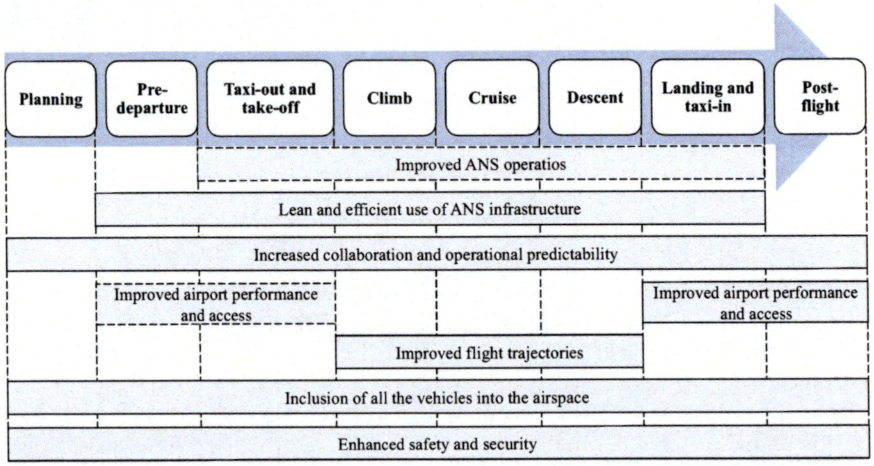

Fig. 1.10 Objectives of SESAR JTU initiative

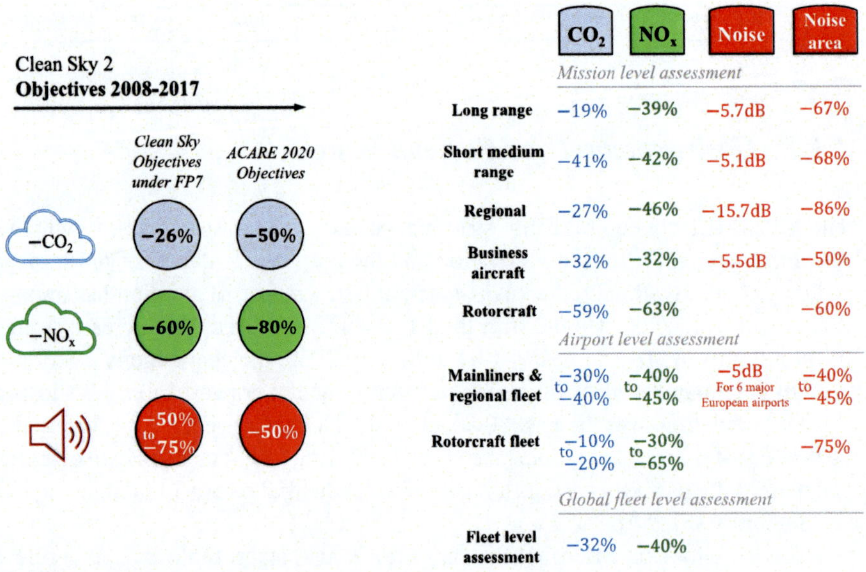

Fig. 1.11 Objectives of the CleanSky JTI program

Society and governments can also benefit from the MEA/MEP concepts, having the opportunity of creating initiatives at the country level, that could include

- job creation and increased employment in highly skilled sectors like aerospace, mechanics, electronics;
- potential long-term boost to productivity;
- enhanced mobility and increase of people quality of life.

Governments should thus set up industrial strategies consisting of tax incentives to attract investments from companies interested in aircraft electrification, and potentially co-invest into public–private partnerships. Governments should promote and facilitate regulations to enable technology demonstrators, and enable the certification of new aircraft concepts [51, 52].

1.1.3 Market Issues

In 2019, the aerospace industry was one of the fastest growing sectors in the world, with need for air travel predicted to continuously grow in the upcoming years [2]. In 2020, the economic outlook dramatically changed due to the SARS-CoV-2 (COVID-19) pandemic, making economic and market forecasts almost impossible while dark clouds gathered on the future of the aerospace industry. The SARS-CoV-2 outbreak caused a general economic collapse both in demand and in production. All the economic indicators, both in Europe and the US, reached a record low in March 2020, with the largest monthly drop since the end of the Second World War [49].

Forecasting the effects of the COVID-19 outbreak in mid-2020 was a challenging task, but something could be learned from the past. In January 2020, when the outbreak was confined to China, IATA analyzed the effects on local Asian-Pacific aviation of previous pandemics showing that, in the past, the airline industry has proven resilient to shocks, including pandemics. Even in the outbreak of SARS, monthly international passenger traffic returned to its pre-crisis level within 9 months [5], Fig. 1.12.

Only a few months later, in May 2020, it was clear that the scenario could have been definitely worst. In [36], an economic forecast including the pandemic effects was presented based on airplane movements extracted from online flight tracking platforms and online booking systems. The prediction was compared with those based only on data up to 2019, also analyzing the global socioeconomic effects of the flight ban in the first quarter of 2020. The conclusion was that that the impact of aviation losses could have negatively reduced World GDP by 0.02– 0.12% according to the observed data and, in the worst-case scenario, at the end of 2020 the loss could be as high as 1.41–1.67%.

The comparison between IATA forecasts at the end of 2019 [2] and mid-2020 [3] are shown in Table 1.4. It is evident that the forecasts have been strongly reviewed and the predicted losses with respect to 2019 could be terrible. In 2020, with respect to 2019, IATA predicted approximately a 50% drop in passengers departures and Revenue Passenger Kilometres (RPKs); about 40% drop in seats and scheduled flights; about 30% drop in global fleet and jobs.

The number of dismantled old airplanes in 2020 was predicted to be four to five times more than in the past with no replacement by new ones. Commercial airlines in May 2020 had around 960 new aircraft scheduled for delivery. This is approximately 40% less than the number originally planned at the beginning of the year, and it was very likely that airlines will have considered further order cancelations. In fact, at the

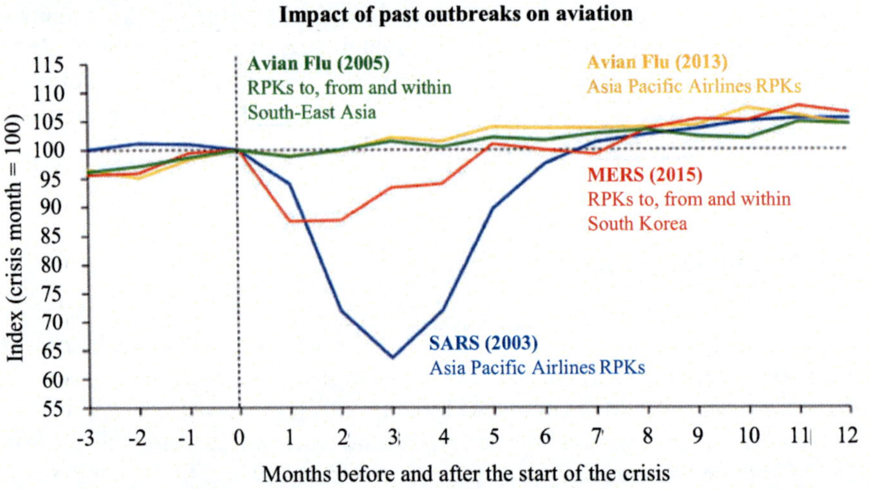

Fig. 1.12 Global RPK (Revenue Passenger Kilometer) values in correspondence of SARS, MERS, and Avian Flu. © International Air Transport Association, 2020. What can we learn from past pandemic episodes? All Rights Reserved. Available on IATA Economics page https://www.iata.org/en/publications/economics/

Table 1.4 Comparison of global standard airline operations indicators in 2019 (first column) with the 2020 forecast made before COVID-19 (second column) and with the 2020 forecast made in June 2020

	2019 data	2020 forecast before COVID-19	2020 forecast after COVID-19
Spend on air transport [$ billion]	876	908	434
Passenger departures [million]	4543	4723	2246
Revenue Passenger Kms [billion]	8680	9038	3929
Available seats [million]	4.5	4.7	2.8
Scheduled flights [million]	38.9	40.3	23.1
Aircraft fleet	29697	31375	20261
Employment [million]	2.79	2.90	1.87

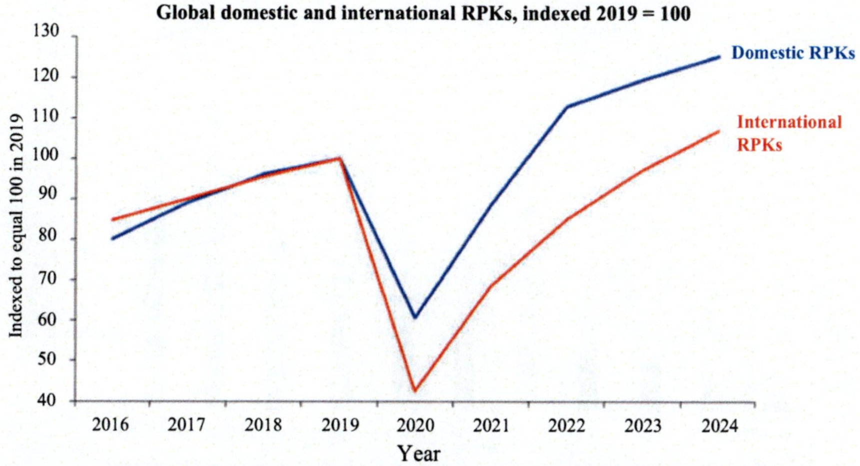

Fig. 1.13 Global RPK (Revenue Passenger Kilometer) values in domestic and international flights. © International Air Transport Association, 2020. Outlook for air travel in the next 5 years. All Rights Reserved. Available on IATA Economics page https://www.iata.org/en/publications/economics/

end of May 2020 only 235 new aircraft had been delivered, well less than usual. Also, air fleet maintenance and repair activities were predicted to be strongly affected.

If forecasting at mid-2020 was more difficult than climbing on vertical ice, the real and much more difficult question was: when the airline industry, and aerospace and defense industry in general, will be back to pre-COVID-19 numbers? The answer depended on a large number of factors, ranging from the evaluation of the real impact of the pandemic on the global economy in general and more specifically on the travel and air cargo industry, to other local and prompt factors as the state aid to airlines in Europe, the amount of contribution, the financial tools used and their long-term effectiveness. In May 2020 [4], IATA made an attempt to predict some indicators, based on the forecast on global economic trends and the result was frightful: a positive trend could be regained early in 2021, but the 2019 values could be reached again only in 2024 as shown in Fig. 1.13, as an example, where the predictions of RPKs are shown.

Nonetheless, the CO_2-emissions of the aerospace industry are always under strict examination and their reduction remains a high priority issue. In 2017, the European Commission (EC) signed the Flight Path 2050 strategy which is not expected to be reviewed early [11]. Besides the challenges in safety and security area, infrastructures, and operations, the strategy also aims to mitigate the impact of air traffic on the environment by agreeing on the following targets:

- CO_2-emissions reduced by 70% (compared to 2000 levels);
- NOx-emissions reduced by 90% (compared to 2000 levels);
- noise minimization.

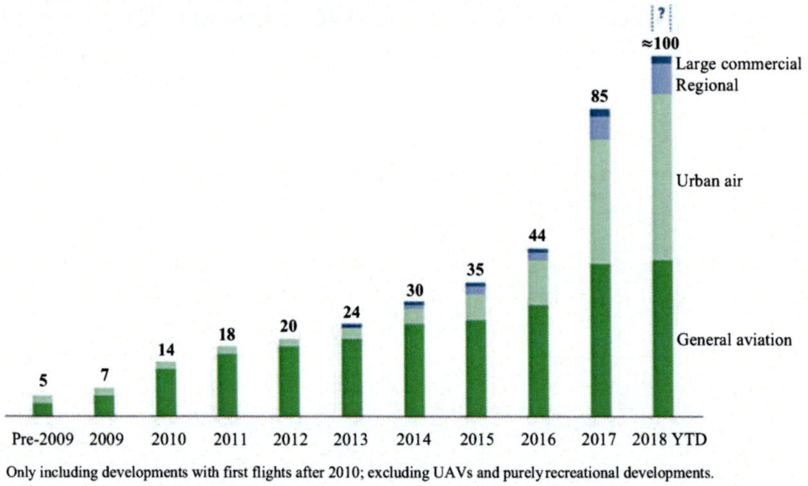

Fig. 1.14 Cumulative number of programs on aircraft electrification launched from 2009. Adapted with permission from Roland Berger. Source: Roland Berger, https://www.rolandberger.com/publications/publication_pdf/roland_berger_aircraft_electrical_propulsion_2.pdf

The MEA/MEP objectives play a key role in this market context. In 2017, about 40 new development programs based on aircraft electrification have been announced, and the trend has been maintained up to now. As depicted in Fig. 1.14, an important contribution to this trend is given by the developments of electric Vertical Take-Off and Landing (VTOL) vehicles as well as urban mobility applications, but the interest in electrification is constantly increasing also in other aviation segments [51].

The most significant barriers to be overcome before achieving the MEA/MEP targets are

- *Technological barriers*, mainly related to
  - limited power density of electric power systems (e.g., batteries and motors for the electric propulsion);
  - thermal management;
  - high-voltage wiring;
  - power electronics reliability.
- *Regulation barriers*, mainly related to
  - novel architecture certification;
  - emissions regulation;
  - air traffic regulation, with particular reference to the urban mobility sector.
- *Market demand barriers*, related to both airlines' operators and passengers, and due to taking on the risk of flying before aircraft electrification has become widespread.

A key factor for the continuation of the steady transition to MEA/MEP concepts is the increase in electric actuation and electrically powered systems. Electric actuators are expected to have a dramatic increase in application for flight controls, landing gears, thrust vectoring systems, with subsequently increased use of components such as MOSFETs, diodes, high-power-density modules. In the sector of large commercial aircraft, most of the current platforms use hydraulic actuation systems. Thus, the increase in the application of electrically powered actuators would result for airframe manufacturers in the substitution of parts/equipments (and, potentially, switching of suppliers), and there would be an increase of the market share for suppliers of electric actuators (with consequent decrease for the suppliers of hydraulic actuators).

Within this context, independently of the COVID-19 crisis, the market increase in the electrical power systems is expected to be substantial in the next 10 years. In [52], multiple sub-scenarios of MEA platforms being introduced into the market have been modeled. One such scenario considered a relatively fast (but still evolutionary) uptake, in which major airframe manufacturers introduce a new MEA by 2025. According to this model, it was found that the total installed electric generation capacity would grow at 8% into 2030, compared to 3% without the introduction of new further electrified platforms.

1.2 Impacts of Research and Development of Electro-Mechanical Actuators

This section reviews the technological solutions toward electrification of the actuation functions, starting from hybrid solutions to all-electric ones. Major focus is clearly given to the electro-mechanical actuators, illustrating their architecture, subsystems, and components.

1.2.1 Electrically Powered Actuators

A key factor for achieving the MEA objectives is the use of electrically powered actuation systems. Different technological solutions have been developed while pursuing the MEA path, i.e.:

- Electro-Hydrostatic Actuators (EHAs), among which
  - Variable-Displacement EHA (EHA-VD) or Integrated Actuator Package (IAP^{TM})
  - Fixed-Displacement EHA (EHA-FD).
- Electro-Backup-Hydrostatic Actuators (EBHAs);
- Electro-Mechanical Actuators (EMAs).

The following sections will review the main concepts of these technologies.

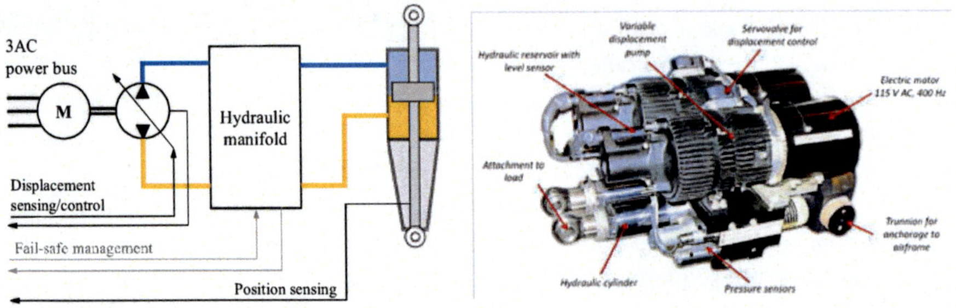

(a) (b)

Fig. 1.15 **a** EHA-VD: working principle; **b** design for the Lockheed C-141 aileron. Reproduced, with permission, from Mare, J.C.: Aerospace Actuators 2: Signal-by-Wire and Power-by-Wire. John Wiley & Sons © ISTE Ltd 2017

1.2.1.1 Variable-Displacement Electro-Hydrostatic Actuator

The EHA-VD or IAP™ concept is based on the idea of using a miniaturized hydraulic plant within the actuator assembly. The EHA-VD integrates a fixed-speed motor, directly connected to the aircraft electric power supply, with a variable-displacement pump, in which the displacement is regulated by a proportional servo valve, Fig. 1.15a.

Figure 1.15b shows a duplex EHA-VD aileron actuator developed and flight tested in the late 1980s on a Lockheed C-141 military transport aircraft [42]. This solution has been initially adopted when the motor power/control electronics appeared to be not sufficiently matured, but the rapid advances in power electronics and the excessive heating of the fluid given by the EHA-VD pump quickly led to abandon this concept.

1.2.1.2 Fixed-Displacement Electro-Hydrostatic Actuator

In an EHA-FD, the power is instead regulated by the electric motor via the control electronics. The EHA-FD integrates a fixed-displacement pump, a variable-speed electric motor, and a Motor Power Electronics (MPE), Fig. 1.16a. The actuator control is performed by closed-loop feedbacks on both motor speed and output position. Flight controls actuated by EHA-FD can be found on Airbus A400M and on JSF F-35 [43], see Fig. 1.16b.

1.2.1.3 Electro-Backup-Hydrostatic Actuator

A third "hybrid" solution is represented by the so-called EBHA, which integrates two actuation technologies in a unique assembly: a conventional Servo-Hydraulic Actuator (SHA), powered by the centralized aircraft hydraulic network and an EHA-FD, Fig. 1.17a. The resulting hybrid actuator, having redundant and dissim-

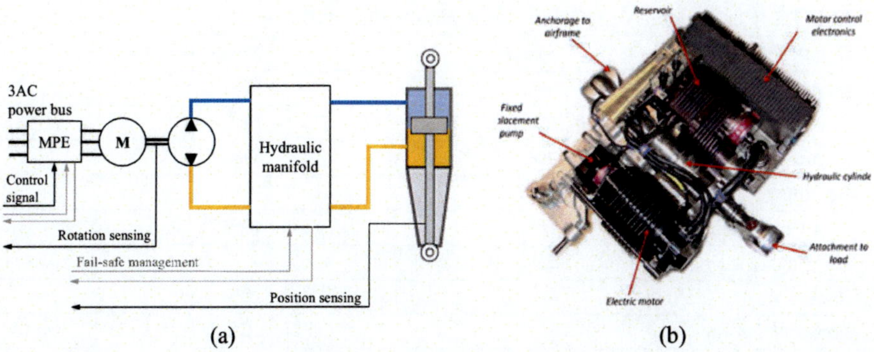

Fig. 1.16 **a** EHA-FD: working principle; **b** JSF F-35. Reproduced, with permission, from Mare, J.C.: Aerospace Actuators 2: Signal-by-Wire and Power-by-Wire. John Wiley & Sons © ISTE Ltd 2017

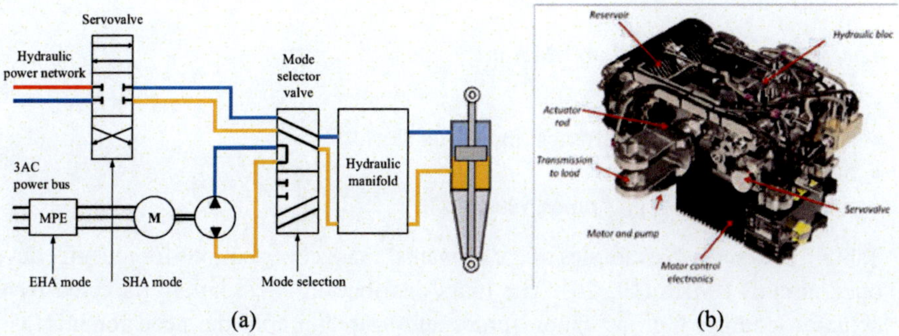

Fig. 1.17 **a** EBHA: working principle; **b** design for the A380 spoiler. Reproduced, with permission, from Mare, J.C.: Aerospace Actuators 2: Signal-by-Wire and Power-by-Wire. John Wiley & Sons © ISTE Ltd 2017

ilar power/control paths, assures high reliability and safety, so that it was introduced by Airbus on the A380 for the actuation of four spoilers and four rudders [58], Fig. 1.17b.

1.2.1.4 Electro-Mechanical Actuator

In EMAs, no hydraulic power fluid is used, and the actuation power flows from the aircraft's electrical supply to the load via a mechanical transmission. In aerospace applications, EMAs have long been used for low-power and non-critical actuation functions, but the MEA trend, together with the constant and relevant increase of the TRL of electric and electronic devices, now allows to consider their application for high-power and safety-critical functions too [48].

Table 1.5 Contributions of different components to the EMA mass

Component	Spoiler	Nose-wheel steering
Motor	34%	18%
Brake	6%	14%
Reducer	9%	6%
Bearing	13%	–
Screw mechanism	16%	40%
Housing	22%	–
Rack	–	10%
Power electronics	–	12%

1.2.2 EMA Technology

The basic components of an EMA are

- electric motor;
- power and control electronics, including sensors;
- mechanical transmission;
- fail-safe devices (e.g., brakes, clutches).

Table 1.5 shows two examples of components' mass contributions for recently developed aircraft EMAs [19, 20]. The mass distribution is essentially affected by the actuator loading: high actuation forces and torques imply the need for increasing the mass of the mechanical transmission and conversely. It is also worth noting that the mass contribution of the power/control electronics ("Housing" and "Power electronics"+"Rack" in Table 1.5) is roughly constant and it covers 20% of the EMA mass. Similarly, the mass contribution of the components providing the mechanical power output ("Motor" + "Screw mechanism" + "Reducer" in Table 1.5) is roughly constant and it covers 60% of the EMA mass.

1.2.2.1 Electric Motors

Electrical machines for aviation must be thermally robust and highly efficient to limit the power losses and the related cooling, being the machine's weight and volume strictly dependent on its energy consumption. Different electrical machine concepts have been evaluated for the MEA application, i.e.:

- Permanent Magnet Synchronous Machine (PMSM);
- Electrically excited Synchronous Machine (ESM);
- Switched Reluctance Machine (SRM);
- Induction Machine (IM).

Table 1.6 Comparison among different electrical machines. (−): disadvantageous, (o) intermediate, (+) advantageous, (++) greatly advantageous. Reproduced from [33]—originally published open access and licensed under CC-BY https://www.mdpi.com/1996-1073/11/2/344/pdf

Key Characteristic	ESM	IM	SRM	PMSM
Rotor losses	−	o	o	++
Stator losses	++	o	o	o
Windage losses	−	o	−	++
Rotor thermal limitations	o	+	++	o
Cooling options	−	o	o	++
Rotor mechanical limitations	−	o	+	++
Torque-to-inertia ratio	o	o	o	++
Compatibility with bearings	−	o	o	++
High-speed capability	−	o	+	++
Short-circuit behavior	−	++	++	−
Machine complexity	o	+	++	+
Current density	−	+	+	+
Power density	−	+	+	+

Table 1.6 shows the different key characteristics of the electrical machine concepts, and the clear result is that the PMSM is the most feasible solution for aircraft applications [33]. Compared with the other machine concepts, the PMSMs are characterized by higher efficiency, higher power density, lower heat production in the rotor, and are capable of sensorless control, although the magnets are comparatively expensive. Different rotor designs are present for PMSM, i.e., V-shaped buried magnets, beam-like buried magnets, surface-mounted magnets, or multilayer buried magnets [33, 53], Fig. 1.18. Each rotor design has its benefits concerning flux density, flux distribution, mechanical stability, or weight, and the solution to be addressed depends on the control application. As an example, buried solutions are preferable for obtaining sinusoidal back-electromotive forces, while surface-mounted solutions are suitable for trapezoidal outputs.

The suitability of a motor is often addressed by the power-to-weight ratio, even if this number can be not significant in some applications, since the power is compared to the overall weight which includes the housing and bearings and the related materials. As these can considerably differ depending on the application, the power-to-volume ratio is in some cases more adequate, as it is not material dependent. An

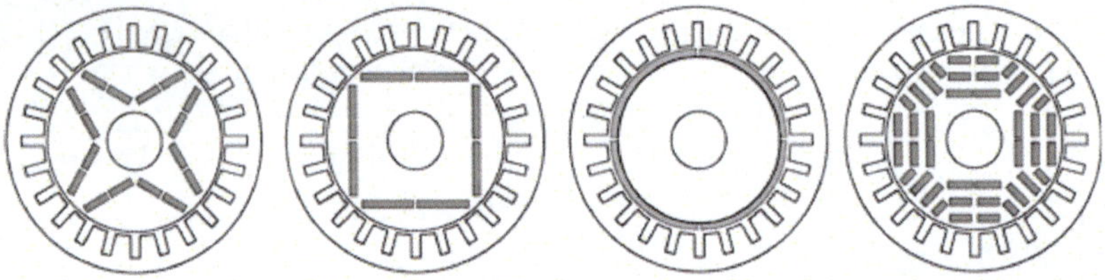

Fig. 1.18 Different rotor concepts of permanent magnet synchronous machines. From left to right: V-shaped buried; beam-like buried; surface mounted; multilayer. Reproduced from [33]—originally published open access and licensed under CC-BY. https://www.mdpi.com/1996-1073/11/2/344/pdf

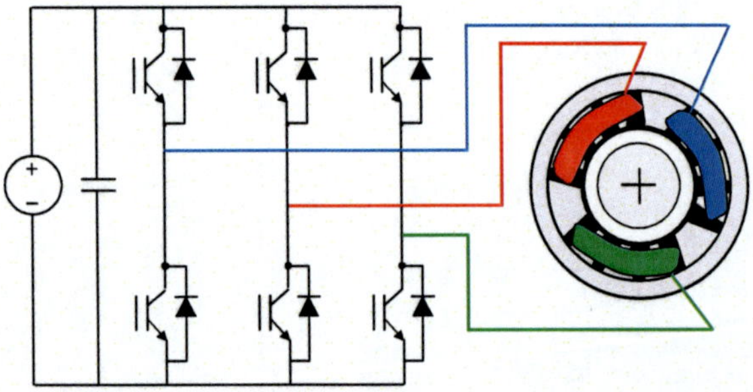

Fig. 1.19 Conventional power electronics architecture with 3-phase motor

alternative choice is the *Esson's number* n_E in (1.2), as it incorporates the rotational speed into the power-to-volume ratio, thus providing a sort of torque-to-volume ratio [17]:

$$n_E = \frac{P_R}{D_b^2 l_{mp} \omega_R}, \tag{1.2}$$

where P_R is the rated power, D_b is the motor bore diameter, l_{mp} is the length of the "active" motor parts, and ω_R is the rated rotational speed.

1.2.2.2 Power and Control Electronics

The EMA control is performed at two levels and by two well-separated sections of its electronic control box:

- the motor drive, performed by the high-power section of the electronic box (briefly, the *power electronics*);
- the actuator closed-loop control and health monitoring, performed by the low-power section of the electronic box (briefly, the *control electronics*).

The power electronics is the section that is responsible for the regulation of the electrical power coming from the aircraft electrical system. This regulation is performed by means of electronic bridges, in which a set of power switches (e.g., MOSFETs) receives commutation commands from the control electronics, Fig. 1.19.

The elaboration and transmission of the commutation commands must satisfy several constraints [42]:

- **Isolation**: the control electronics operates at low voltage, and it must be isolated from the power section, because it contains sensitive components (processors, etc.). The isolation is typically achieved by transformers with a high-frequency carrier or by optocouplers.
- **Conditioning and amplification**: the commutation commands have to be conditioned and amplified in order to adapt them to the power switches needs.
- **Exclusivity**: the commutation commands must assure that no simultaneous closing of both power switches on a leg is possible, as this would short-circuit the DC aircraft power supply.

Different approaches are used to implement the motor drive, i.e., to perform the on-off activation of the power switches. Among them, two techniques are mainly used for aircraft EMAs:

- **Six-step control**: this technique is used to drive the so-called BrushLess DC Machines (BLDCMs), a class of PMSMs having trapezoidal back-electromotive forces. It is an electronic version of the mechanical commutation of brushed DC machines, Fig. 1.20. The control is easy to implement, because it only needs discrete data about the rotor angle (i.e., 60° sector of the electrical cycle), so that the stator can simply be fitted with discrete sensors (e.g., Hall-effect sensors), which detect the presence of the magnetic field generated by the rotor magnets. In addition, the six-step control requires no calculation since the commutation commands are a combinatorial function of discrete signals coming from these sensors [22]. As drawbacks, the torque ripple increases at high speeds [42].
- **Field-oriented control** (FOC, or *vector control*): The FOC technique limits the downsides of the six-step control by generating sinusoidal phase voltages that have no fronts. High performances can be achieved with BrushLess AC machines (BLACMs), a class of PMSMs having sinusoidal back-electromotive forces. As a drawback, the complexity of the control electronics is higher (since significant real-time signal processing and computation is needed), and the control also requires a very precise measurement of the rotor angle.

Concerning the control electronics, the functions and the algorithms to be implemented in the closed-loop control laws of an EMA depend on the actuator application, i.e.:

- *position-controlled* EMAs are typically used for primary flight surfaces, nose-wheel steering, and thrust vectoring control;
- *speed-controlled* EMAs are used for landing gears extension/retraction as well as secondary flight surfaces (flaps, slats, airbrakes, etc.);
- *force/torque-controlled* EMAs are suitable for brakes or force-feedback pilot inceptors.

Figure 1.21 provides an example of the closed-loop control architecture for a position-controlled EMA with FOC-based motor driver.

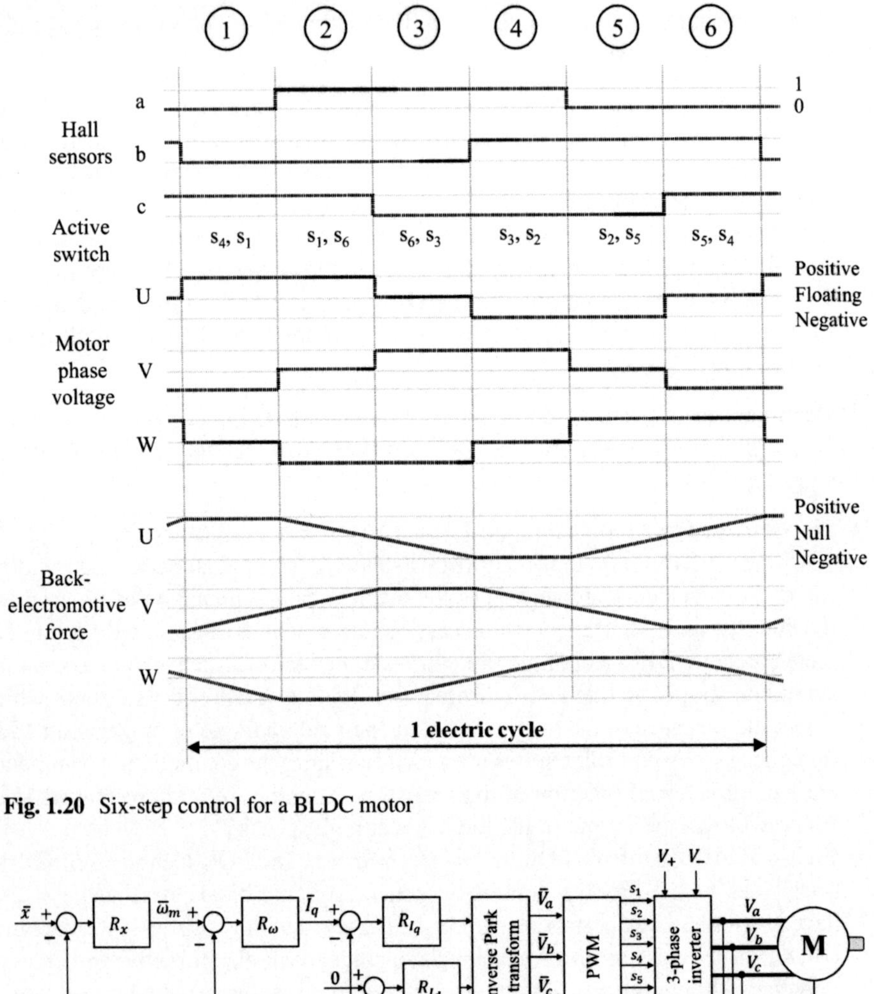

Fig. 1.20 Six-step control for a BLDC motor

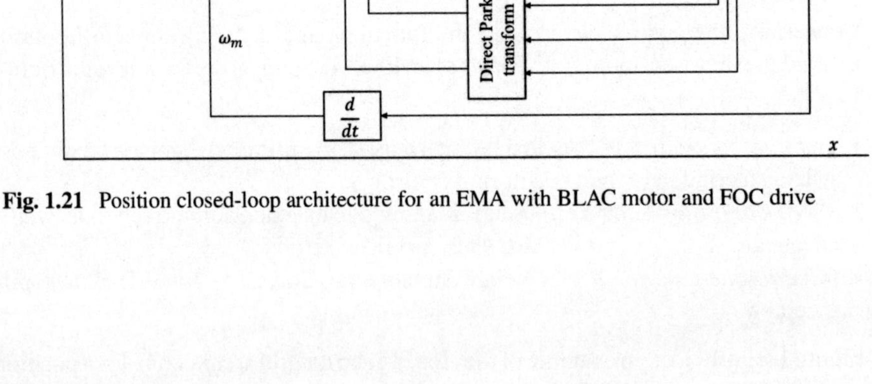

Fig. 1.21 Position closed-loop architecture for an EMA with BLAC motor and FOC drive

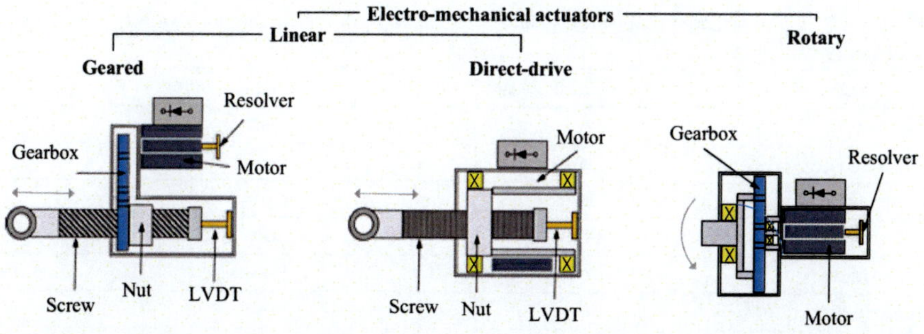

Fig. 1.22 Configurations of mechanical transmission for EMAs. © 2018 SAGE Publishing. Reprinted, with permission, from [48]

1.2.2.3 Mechanical Transmission

The first basic issue to be addressed when defining the mechanical transmission of an EMA is to decide if *linear* or *rotation* output is to be given. Traditionally, the actuation of primary flight control surfaces and landing gears extension/retraction was obtained by linear actuators. One of the main reasons for adopting this configuration was related to the use of hydraulic actuators. Rotary hydraulic actuators typically have large losses, so that linear output is undoubtedly preferable. The transition from hydraulic to electro-mechanical technology could in principle lead to actuators with rotating output, but the integration needs of the novel solutions into the existing architectures still lead to consider linear actuators as the reference solution.

Aircraft actuation functions are characterized by mechanical power with high force and low speed, so the use of an electric motor directly connected to the load is typically not possible due to the constraints of weight and size. It is therefore essential to integrate into the EMA mechanical transmission low-pitch nut screws (if linear actuation is used) or high-ratio gearboxes. Depending on the presence or not of gearboxes, EMAs can be categorized into (Fig. 1.22):

- *Geared EMA*, with the gearbox mounted *in-line* or *off-line* the motor axis;
- *Direct-Drive EMA*, which in some cases integrates the rotor magnets into the rotating element of the screw mechanism.

The preferred configuration for mechanical transmission on EMAs is the *direct-drive* one, especially if the magnets are bonded onto either the nut or the screw shaft [48]. The main advantages are

- low reflected inertia from the motor rotor to the load;
- reduced number of components;
- higher efficiency;
- thermal stability;
- vibration modes at higher frequencies;
- reduced maintenance;

Table 1.7 Comparison of ball screws and roller screws. (o) intermediate, (+) advantageous, (++) greatly advantageous

Criterion	Ball screw	Roller screw
Load	o	+
Life	o	+
Speed	+	
Backlash	+	o
Stiffness	o	++
Reduction ratio	o	+
Efficiency	++	o
Maintenance	o	o

- reduced noise;
- increased accuracy;
- reduced irreversibility load;
- lower backlash.

Concerning the technological solution for the screw mechanisms, both ball screw and roller screw mechanisms can be suitable for aircraft EMAs. The ball screw has lower friction, but lower load capacity. Conversely, the roller screw has higher load capacity but higher friction. Table 1.7 provides a qualitative comparison of the main performance characteristics of ball screws and roller screws.

As a drawback for linear direct-drive EMAs, the screw jacks are susceptible to jamming, particularly when operating under vibration and dynamic loads, so many research efforts have been made on developing and testing the screw mechanisms [14, 37, 38].

1.2.2.4 Fail-Safe Devices

A critical issue to be addressed in the development of aircraft EMAs is the management of the fail-safe mode of the system. In EHAs and EBHAs, these protection functions were effectively and efficiently accomplished via hydraulic components (by-pass valves, pressure relief valves, etc.), while in EMAs they must be implemented by mechanical, electromagnetic, or electric devices. The basic functions to be implemented by *fail-safe* modes are

- **Load limiting**: the capability of maintaining the actuator loads at a fixed and limited value (typically neutral);
- **Load disconnection (anti-jamming)**: the capability of disengaging the blocked elements in the mechanical transmission and reverting safe operating configuration.

The *load limiting* function can be implemented by either passive or active devices. Examples of passive devices are skewed roller no-back brakes, in which the friction torque generated by the brake is proportional to the torque derived from the external load, and they are used when a back-driving load must be counteracted (e.g., brakes). The combination of a skewed roller no-back and a one-way free-wheel clutch (or Sprag clutch) is used in some applications since the device provides a friction torque in the back-driving condition and no torque when the actuator operates against the external load. Since they are dissipative devices, energy regeneration is not possible in back-driving conditions. On the other hand, they do not need any input to engage and disengage. Active brakes are typically based on the use of a solenoid that allows some friction/teethed disks to engage the mechanical transmission and to hold the external load. There are different arrangements for both the solenoid and the disks. The brakes can be power-on, power-off (i.e. that inhibit the motion when energized or de-energized, respectively), and bistable type. Monitoring of the disks' engagement/disengagement can be realized through proximity sensors.

Concerning the *load disconnection* function, the mechanical transmission jamming is surely the most feared event since the consequences can be catastrophic if the failure is not compensated. This is the basic reason why EMAs have been rarely used for safety-critical functions, but the technological context has been changed and the use of EMAs nowadays appears feasible. To implement an anti-jamming function, the mechanical transmission must be disconnected from the external load, but for many aircraft applications, it is not acceptable to have a free-wheeling load motion. For this reason, redundant sections have to be designed and integrated to obtain a jamming-free EMA, which, after the failure, keeps partial or full operability [12, 25, 41, 59].

1.2.3 EMA Research

In Europe, the R&D funding efforts toward MEA/MEP concepts concretely started in the 6th Framework Program (FP6), in which the MOET initiative, coordinated by Airbus (France), was funded for 66.6 MEuro. After that, the EC funding effort towards aircraft electrification has been not only maintained but strongly powered: if in the FP6 the whole MEA/MEP topics were funded for 66.6 MEuro, in the FP7 and in Horizon 2020 the sole funding for actuation systems development achieved 66.3 and 39.7 MEuro, respectively.

R&D projects funded within 7th Framework Program.

The list of R&D projects funded by EC within the FP7 on actuation systems development is reported in Table 1.8. It is worth noting that

- the 85% of total funding was dedicated to electro-mechanical actuation;
- the rest of funding was assigned to projects on servo-hydraulic actuation;
- no funding was given to projects on electro-hydrostatic actuation, which were thoroughly investigated within FP6.

Table 1.8 R&D projects on actuation systems developments funded within FP7

Acronym	Topic	kEuro	Coordinator
CREAM	Compact and Reliable EMAs	6373	Safran (FRA)
HP-SMART EMA	High-power-density EMA for gas turbine control	1917	SENER (SPA)
ELETAD	Electrical Tail Drive	2478	UniBristol (UK)
ACTUATION2015	Modular EMAs for aircraft and helicopters	34374	Goodrich (FRA)
EMAS	Electric motor and sensor design	190	Anotato (GRE)
RETAX	EMA for rotorcraft landing gears	945	Michelin (SUI)
ELTESTSYS	Electrical test bench drive systems	646	STRAERO (ROM)
AEGART	A/C power converters	1618	UniNottingham (UK)
SafePEM	Reliable power bus design	497	TTTech (AUT)
ARMLIGHT	EMA for main landing gear	748	CESA (SPA)
E-SEMA	EMA for gas turbine control	968	SENER (SPA)
HPEM	High performance electric motors	249	Anotato (GRE)
MAGBOX	Aeronautical Magnetic gearbox	248	UniMadrid (SPA)
ROTOPOWER	Power converters for swashplate EMAs	345	Castlet (UK)
DREADS	Drivetrain rolling elements	1034	UmbraGroup (IT)
FASTDISC	Disconnecting device for jam-tolerant EMA	832	UniStuttgart (DEU)
HOLMES	EMA health monitoring	489	UmbraGroup
FLIGHT-EMA	EMA and ECU for FCS	1151	CESA (SPA)
TESTHEMAS	Test rig for helicopter EMAs	911	Tecnalia (SPA)
ORPASV	EHSV for open rotor pitch actuator	3744	Zodiac (FRA)
HYPSTAIR	Hybrid propulsion systems	6551	Pipistrel (CZE)

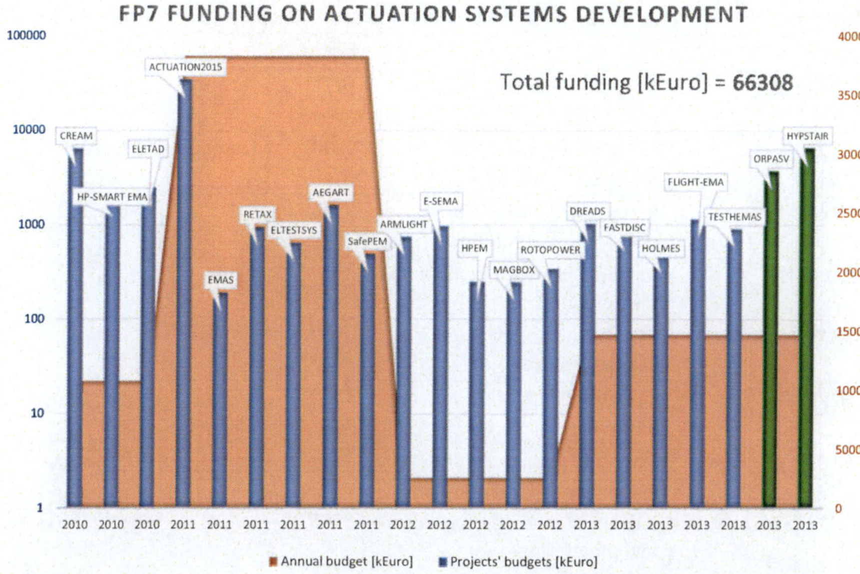

Fig. 1.23 Overview on R&D projects funded in FP7 (in blue: EMA topics; in green: SHA topics)

It can be noted from Fig. 1.23 that 21 R&D projects were funded with a total effort of 66.3 MEuro. The funding rate was about 15 MEuro/year, with a peak in 2011. With reference to EMA developments only, and by excluding the high-funded projects ACTUATION2015 and CREAM, each R&D project approximately received 1 MEuro funding.

R&D projects funded within Horizon 2020.

The list of R&D projects funded by EC within Horizon 2020 on actuation systems development is reported in Table 1.9. In this case, the 97% of total funding was dedicated to electro-mechanical actuation. It can be noted from Fig. 1.24 that 20 R&D projects have been funded with a total effort of 39.7 MEuro. The funding rate has been of about 6 MEuro/year, with a peak in 2015. With reference to EMA developments, each R&D project approximately received 2 MEuro/year funding.

1.3 State of the Art of Aircraft EMA Technologies

This section reviews recent developments in the employment of EMAs for aerospace applications such as flight controls, landing gears, nose-wheel steering, thrust vector control, brakes, and more innovative functions.

Table 1.9 R&D projects on actuation systems developments funded within Horizon 2020

Acronym	Topic	kEuro	Coordinator
ALLEGRA	Advanced low noise landing gear EMA for regional aircraft	1971	UniDublin (IRE)
ASTIB	Flight simulator and iron bird for regional aircraft	5000	INSA (FRA)
COSTAR	Innovative actuators for compound fast rotorcraft	2500	Protom (IT)
LG EMA	Landing gear for a compound fast rotorcraft	3500	–
FBW EMA	FBW actuators for tiltrotor primary controls	3500	–
BK EMA	Compact EMA brake for small aircraft	4500	–
REPRISE	EMA for primary controls with advanced health monitoring	1309	UmbraGroup (IT)
ALGESMO	Advanced landing gear sensing and monitoring	3555	Meggitt (UK)
ISSELUB	Innovative technologies for lubricated elements	681	Tekniker (SPA)
EMA4FLIGHT	EMA and ECU for FCS 1857 Tecnalia (SPA)		
VALEMA	Verification and Validation tests of EMA and ECU at TRL 6	1092	GMV (SPA)
TAIRA	Fault-tolerant aileron EMA	1709	Honeywell (CZE)
HYDRORIG 3G	Hydraulic rig for actuator integration	330	AIERS (SPA)
AMuLET	Advanced ECU for morphing leading edge	623	SkyLife (SPA)
FluidER	Sensors for hydraulic fluid monitoring	561	Tekniker (SPA)
TAILTEST	Test rig of innovative rotorcraft vertical tail	697	Vyzkumny (CZE)
SMAR-TeR	Smart active inceptors for tilt rotor	1309	Mare Engineering (IT)
LG EMA	EMA for landing gear system for small-air transport	600	–
TR EMA	Innovative thrust reverser EMA	900	–
PC EMA	Advanced pitch control EMA	3500	–

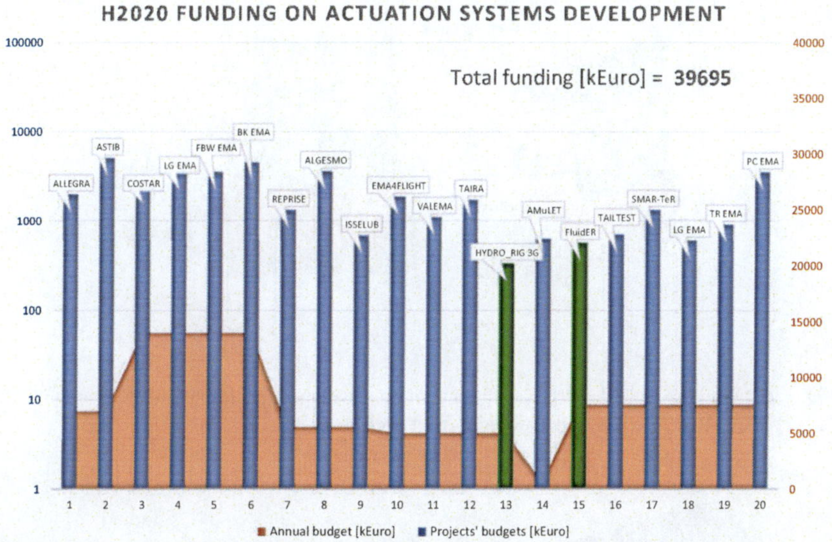

Fig. 1.24 Overview on R&D projects funded in Horizon 2020 (in blue: EMA topics; in green: SHA topics)

1.3.1 Flight Controls

As a result of the combination between the strict reliability/safety requirements for airworthiness certification and the current levels of technology, the actuation systems of flight controls typically have redundant components/subsystems, in order to maintain the operability even if one or more faults occur, a feature which is often defined *fault-tolerance capability*.

The redundancy can be implemented either at actuator level by using nonredundant actuators (*simplex* EMAs) working in parallel, or at subsystem level with EMAs including redundant components (*fault-tolerant* EMAs) [12, 15, 16, 23]. In any case, EMAs for flight controls must implement a *fail-safe capability*.

1.3.1.1 Simplex *Fail-Safe* EMA

An example of simplex EMA is given by the actuator developed by UmbraGroup (Italy) through recent R&D activities [44, 45, 47] (in particular this actuator was used in the first phase of the REPRISE project, see Sect. 4.2.2). The EMA (Fig. 1.25) has a linear output, and the electro-mechanical power conversion is obtained by a direct-drive BLDCM coupled with a ball screw assembly. The system is characterized by a very compact design, and it is made of

- three-phase BLDCM;
- direct-drive mechanical transmission with ball screw assembly (Fig. 1.26a);

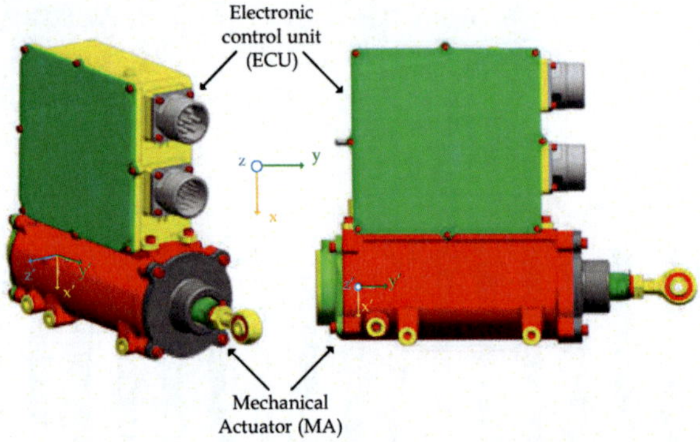

Fig. 1.25 Simplex EMA for primary flight controls developed by UmbraGroup: general layout. © [2017] IEEE. Reprinted, with permission, from [44]

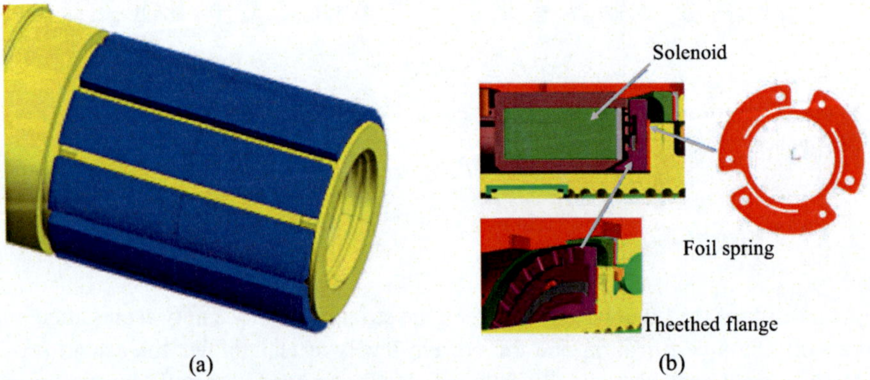

Fig. 1.26 Simplex EMA for primary flight controls developed by UmbraGroup: **a** magnets mounted on the rotating nut; **b** electromagnetic brake. *Source* courtesy of UmbraGroup

- ECU box assembled with the actuator cylinder;
- electromagnetic safety brake (Fig. 1.26b);
- simplex sensing system, based on three current sensors and one LVDT position transducer.

The EMA integrates an electromagnetic power-on brake with teethed flange, which is used to implement the *load holding capability*. Thus, the brake safely stops the screw nut (i.e., the output rod) in case of failure or to hold it at the initial system activation. With no voltage applied to the brake, the teethed flange, held by a foil spring that is attached to the ball nut, rotates together with the motor rotor and the output rod is capable to move. When the solenoid is energized, the magnetic attraction generated on the teethed flange overcomes the foil spring force, causing the engagement of the tooth on the brake housing and the ball nut lock, so that no movement of the output rod is allowed.

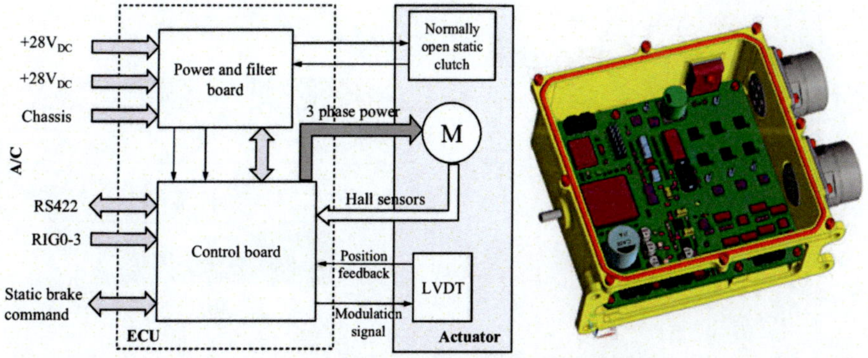

Fig. 1.27 Simplex EMA for primary flight controls developed by UmbraGroup: ECU box. *Source* courtesy of UmbraGroup

The EMA ECU (Fig. 1.27) has two interface connections: one for the power supply lines and chassis from aircraft, and the other for RS422 lines, rigging lines and static brake discrete commands. The ECU basically consists of two boards:

- *Power and Filter Board*: used to implement filtering and protection capabilities for the Control Board 28 VDC electrical supply, as well as to command the electromagnetic brake.
- *Control Board*: used to implement the control and monitoring algorithms, to drive the motor inverter, as well as to communicate with flight control computers via two RS422.

In particular, the Control Board basic functions are

- supply for the motor and low-power devices;
- LVDT conditioning and acquisition;
- execution of digital closed-loop control algorithms;
- execution of monitoring algorithms;
- RS422 communication;
- PWM drive via six-step control;
- Hall effect sensors interface;
- brake control.

1.3.1.2 Redundant *Fault-Tolerant* EMA

A relevant example of fault-tolerant EMA is given by another recent development by UmbraGroup (Italy). In this solution, two independent direct-drive BLACMs, equipped with dedicated brakes, engage an intermediate element (screw shaft), which has three threaded portions: two ones on the external diameter for the motors' ball-nuts, and one on the inner diameter for the output shaft, Fig. 1.28. The kinematics thus allows different possible operating modes [25]:

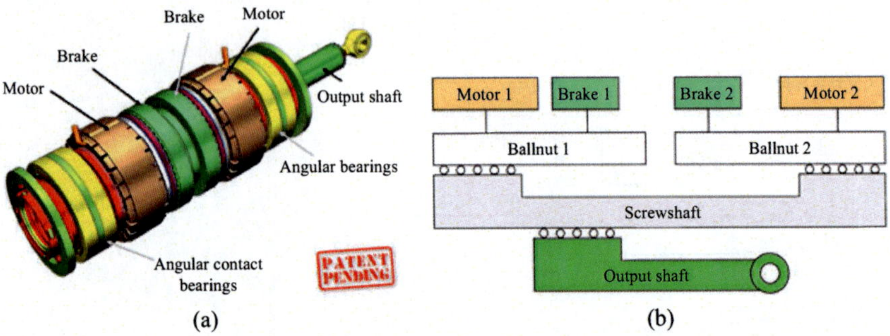

Fig. 1.28 Fault-tolerant EMA developed by UmbraGroup: **a** internal layout; **b** basic kinematic concept. *Source* courtesy of UmbraGroup

- *Active–active* modes, in which the two motors rotate to let the screw shaft rotate together with the motors' ball-nuts, by generating the translation of the output shaft;
- *Active–standby* modes, in which only one motor rotates and the other is held by the brake, so that the screw shaft has a roto-translating motion, and the output shaft translates.

The architecture of the closed-loop control and monitoring system of the fault-tolerant EMA is depicted Fig. 1.29. The sensors' system includes

- n. 2 current sensors for each motor phase;
- n. 2 resolvers for each motor;
- n. 2 cone-type proximity sensors, to measure the screw shaft position;
- n. 2 LVDT transducers.

This solution is used in the second phase of the REPRISE project, see Sect. 4.2.3.

1.3.1.3 EMA Developments for the A320 Aileron

Other relevant examples of EMA developments for flight control applications were gained from the results of a number of projects, carried out in the 2000s and 2010s, concerning the application of the more electric actuation for the A320 aileron.

In particular, in the MOET program, Goodrich developed a linear direct-drive simplex EMA with remote electronics, where a particular focus was placed on the roller screw endurance tests in order to improve the models used for service life calculations and to develop condition monitoring algorithms. Sagem developed a TRL6 linear EMA with integrated electronics, while UTC focused on enhancing the modularization and standardization of EMA elements for cost reduction purposes [42].

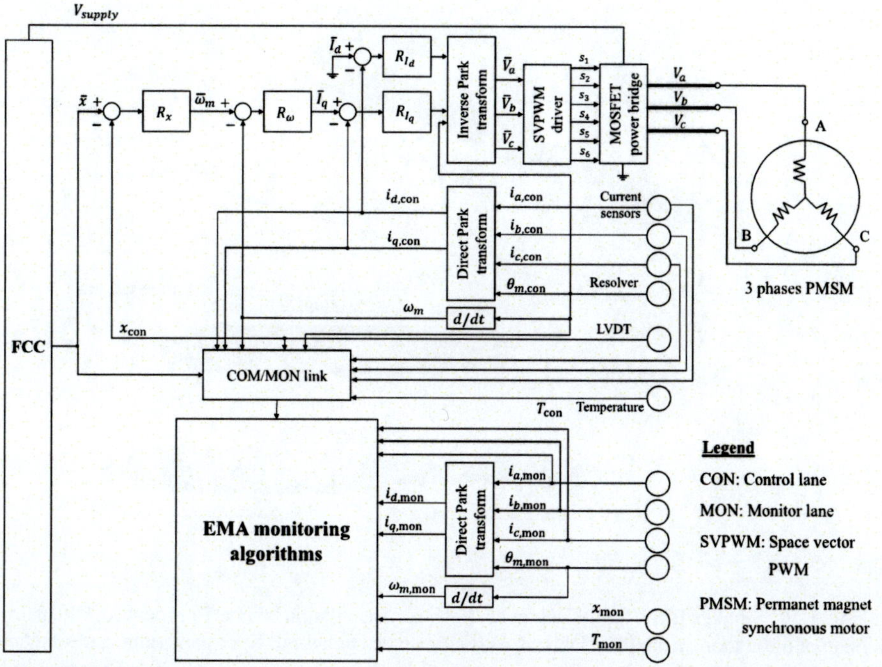

Fig. 1.29 Fault-tolerant EMA developed by UmbraGroup: control and monitoring system. *Source* courtesy of UmbraGroup

1.3.2 Landing Gears

The main difficulty encountered when using EMA for extension/retraction is related to the implementation of the *fail-safe mode*, which must allow a damped free-fall of the landing gear in case of loss of the actuation function. For this reason, EMAs for landing gears extension/retraction are typically simplex solutions, with also simpler closed-loop control algorithms with respect to flight controls (speed control is typically sufficient).

Several research programs in Europe have focused on extension/retraction actuated by EMA, particularly for the evaluation and testing of anti-jamming solutions. Some relevant results of EMA developments for landing gears are reported in [42]. A jamming-tolerant EMA was developed in the ARMLIGHT project, which uses a secondary rod inside the screw for transmitting motion to the load, and, in case of nut–screw jamming, the internal rod is separated from the screw. A direct-drive EMA was developed in the French-funded project MELANY, where the electric motor moves two nested roller screws and an electromagnetic damping device is used to control the free-fall motion. In the CISACS project, a direct-drive EMA with roller screw had a system for releasing the axial thrust bearing of the nut, and a screw-integrated hydraulic piston to provide damping for the *free-fall* motion.

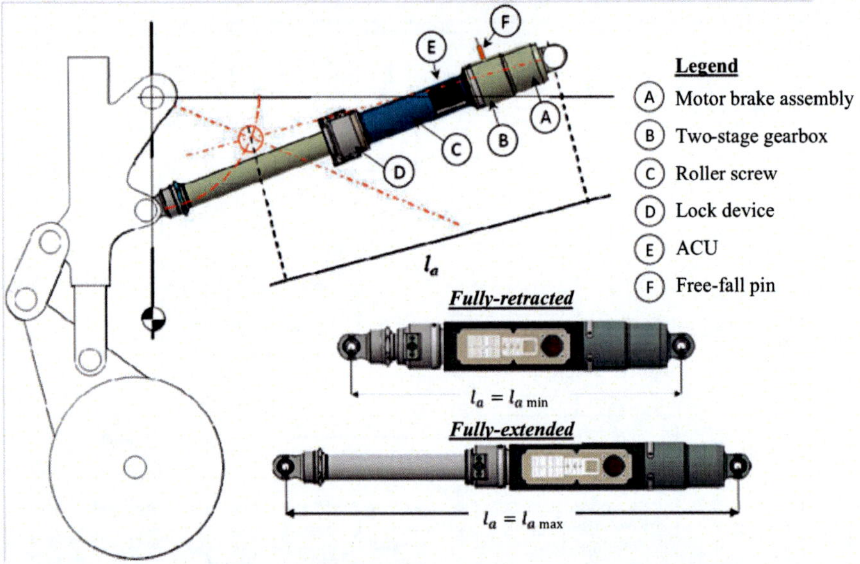

Fig. 1.30 Simplex EMA for helicopter landing gear: installation layout. Reproduced, with permission of the authors, from [26]: Di Rito, G., Galatolo, R., Schettini, F.: Experimental and simulation study of the dynamics of an electro-mechanical landing gear actuator. In: 30th Congress of the International Council of the Aeronautical Sciences (ICAS). Daejeon, South Korea, 2016

Another relevant example is given in Fig. 1.30, by the geared EMA for helicopter landing gear extension/retraction developed by Mecaer Aviation Group (Italy) within the Italian-funded project "Industria2015" [26]. The speed-controlled EMA is essentially composed of

- three-phase BLDCM;
- electromagnetic power-off brake;
- mechanical transmission with two-stage gearbox and low-pitch planetary roller screw;
- mechanically driven lock device for holding the EMA in fully extended position;
- ECU box implementing the six-step motor drive, the closed-loop control and monitoring algorithms, and the fail-safe mode management;
- mechanically driven load disconnection device for the free-fall extension in emergency conditions (free-fall pin), which allows the free motion by manually removing a pre-loaded pin connecting the annular gears of the two stages of the gearbox.

The EMA behavior has been characterized at a high level of detail, in order to validate the dynamic models of the actuator, by using a real-time hardware-in-the-loop testing system available at the University of Pisa (Italy) Fig. 1.31, in which the EMA performances have been evaluated at different operating conditions in terms of voltage supply and mechanical loadings.

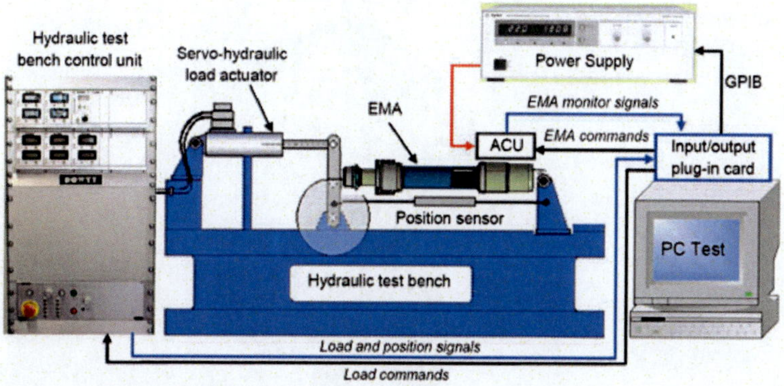

(a)

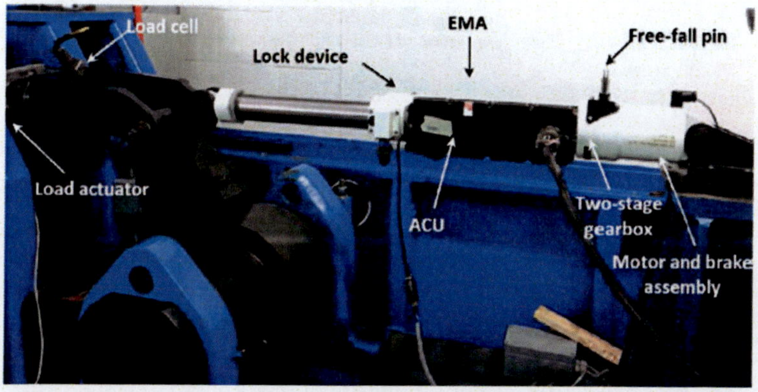

(b)

Fig. 1.31 Simplex EMA for helicopter landing gear: **a** test system; **b** rigging. Reproduced, with permission of the authors, from [26]: Di Rito, G., Galatolo, R., Schettini, F.: Experimental and simulation study of the dynamics of an electro-mechanical landing gear actuator. In: 30th Congress of the International Council of the Aeronautical Sciences (ICAS). Daejeon, South Korea, 2016

1.3.3 Nose-Wheel Steering

Complementarily to the landing gear extension/retraction case, the actuation of the nose-wheel steering has the necessity to a wheel free-castoring in case of actuator fault. This motion allows the wheel to self-align during its rotation, without exerting significant tire loads, so that the aircraft direction can be controlled via other commands (rudder or differential brakes). In the free-castoring mode, it is also crucial to generate damping forces capable of avoiding "shimmy" phenomena. In addition, the closed-loop control functions for the nose-wheel steering EMAs are similar to the flight controls' case, so that the EMAs must be position-controlled.

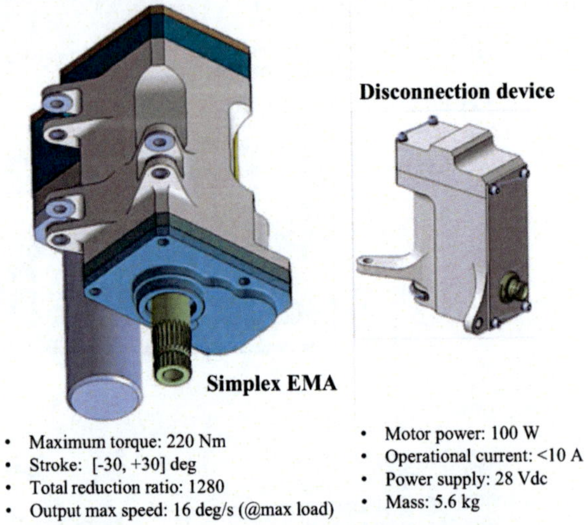

Fig. 1.32 Simplex EMA for nose-wheel steering. *Source* courtesy of Mecaer Aviation Group

An example for the nose-wheel steering application is represented by the geared EMA developed by Mecaer Aviation Group (Italy), Fig. 1.32, where the EMA is made of

- a simplex off-line geared EMA with rotating output with
  - three-phase BLDCM;
  - electromagnetic power-off clutch used to allow the motor shaft free-wheeling (i.e., nose-wheel free-castoring) in case of motor fault;
  - mechanical transmission made of a three-stage gearbox: the first planetary stage is embedded with the motor assembly, the second is a spur stage, and the third one is a multiple planetary stage;
- an electrically actuated disconnection device, capable of disengaging the EMA output shaft from the wheel in case of EMA jamming;
- ECU box implementing the FOC motor drive, the closed-loop control and monitoring algorithms, and the fail-safe mode management.

1.3.4 Brakes

The main design issues related to the aircraft brakes actuation are related to the heat generation. In the 1 disc-brake solution, the brake actuator applies a force to compress a set of discs that are alternatively linked in rotation with the wheel and the landing gear leg respectively, and the decelleration 1111 is transmitted to the tire–runway interface through friction between discs. The kinetic energy to be dissipated can exceed 100 MJ, with an instantaneous power above 5 MW [42]. The heat is initially

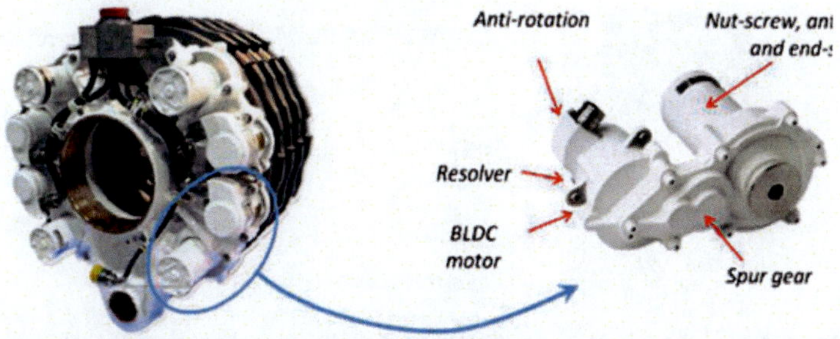

Fig. 1.33 B787 brake EMA by Safran. Reproduced, with permission, from Mare, J.C.: Aerospace Actuators 2: Signal-by-Wire and Power-by-Wire. John Wiley & Sons © ISTE Ltd 2017

stored in the discs' pack (with temperature rise up to 1000 °C in tens of seconds), and then it is slowly released to surrounding parts and environment.

Another key feature of the brake actuation is the need for fast force responses. Actually, when the braking demand is given, the pistons must rapidly get contact with the discs' pack and then modulate the compressing force with a high bandwidth response to implement the anti-skid capabilities.

The application of the MEA concept for brakes can imply enormous advantages in terms of both maintenance costs and dynamic performances, since brake EMAs can be treated as Line Replacement Unit (LRU) elements and enhanced performances can be achieved by implementing closed-loop control algorithms that track a current demand. Figure 1.33 shows a relevant example of EMA developed by Safran for the B787 brake. The solution is based on a simplex geared linear EMA, integrating a BLDCM, a gearbox, and a nut–screw mechanism.

1.3.5 Thrust Vectoring Control

Thrust Vectoring Control (TVC) actuators, mainly developed for space applications, are used to steer the nozzle of an engine, in order to control the thrust vector direction. The interest in EMAs for thrust vectoring functions has been constant over the decades, due to the potentially superior characteristics in terms of dynamic response and reduced maintenance with respect to the SHA solution. Nevertheless, reliability issues slew down the EMA application up to 90 s, when NASA evaluated the replacement of SHAs with EMAs for the TVC of the main engines of the Space Shuttle [30].

Nowadays, several launchers use EMAs for the thrust vectoring, as in the ATLAS Centaur [32] or in the European VEGA launchers [24], Fig. 1.34. Thrust vectoring EMAs are also used in the missile sector, as for in the French strategic missile M51, in which a hollow and concentric electric motor directly drives the nut of an inverted roller screw [42]. In the VEGA launcher, each stage features two EMAs,

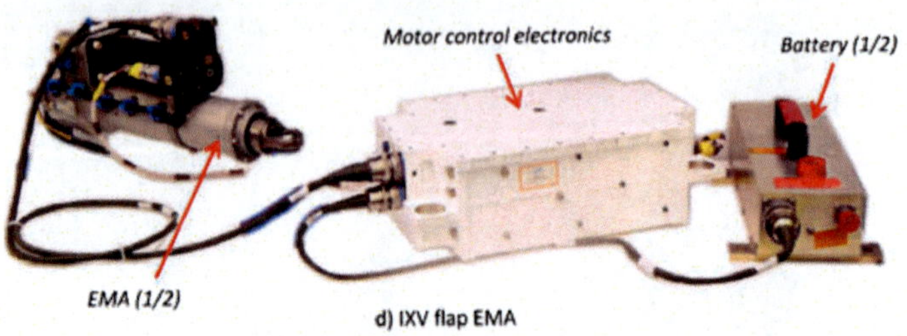

Fig. 1.34 EMA developments for the TVC of the VEGA launcher. Reproduced, with permission, from Mare, J.C.: Aerospace Actuators 2: Signal-by-Wire and Power-by-Wire. John Wiley & Sons © ISTE Ltd 2017

for yaw and pitch, an ECU driving the two motors, lithium-ion power batteries, and the electrical harness connecting these elements. Each simplex position-controlled EMA comprises a PMSM, a parallel axes gear reducer, a nut–screw transmission and the sensing system (motor resolver, LVDT, force sensors, and temperature sensors). Roller screws are used for the first three stages, while the last stage of the launcher uses a ball screw assembly.

1.3.6 Innovative Functions

As a result of the simplification of integration activities for aircraft actuators, the potentialities of innovative functions implying the installation of EMAs in different locations of the airframe are widespread. In particular, we review applications of EMAs to winglet movables and wheel control.

1.3.6.1 Winglet Movables

A relevant example of innovative EMA application is given by the R&D activities carried out within the CleanSky program for the evaluation of novel wing actuation capabilities, aiming to obtain mission-optimized aerodynamic performances, noise abatements, and wing load control and alleviation [1].

In this context, the application of EMAs for winglet movables in the CleanSky2 demonstrator [27] has been analyzed and demonstrated to be feasible. The winglets' actuation system is composed of two independent command chains, each one comprising an ECU box, a simplex direct-drive EMA with linear output, and a movable section, Fig. 1.35. The position-controlled EMA is essentially made of

- three-phase brushless BLDCM with FOC drive;
- "direct-drive" ball screw integrating the motor magnets in the rotating ball nut;
- electromagnetic power-off brake;

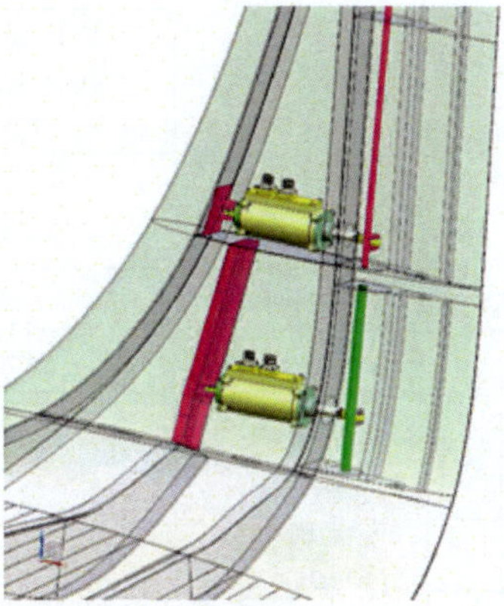

(a)

(b)

Fig. 1.35 Simplex EMA for winglet actuation system developed by UmbraGroup: **a** installation of EMAs; **b** external and internal layout of the actuator. Reproduced from [27]–originally published open access and licensed under CC-BY 4.0. https://www.mdpi.com/2076-0825/8/2/42

- simplex sensing system, with three current sensors, motor resolver, and LVDT position transducer;
- ECU box, including separated modules for power supply, computation, and PWM motor drive functions.

1.3.6.2 Wheel Control

The application of EMA to the control of wheels is essentially focused on enabling the aircraft taxiing without using the main engines or external devices such as tractors or tugs. There is currently a strong interest in these applications, because they are expected to have a relevant impact on "green objectives" for the aerospace sector, mainly in terms of reduction of noise and pollution at airports' neighborhoods [39]. Actually, main engines are optimized for cruising speed, and they are highly inefficient when used in idle mode. Taxiing is thus addressed as one of the biggest contributors to the pollution and noise at the airports (e.g., more than 56% of the total NOx generation in 2002 at Heathrow airport derived from taxiing phases [28]). In

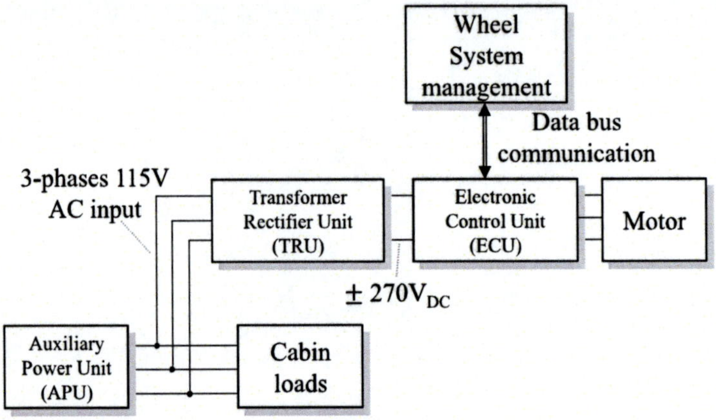

Fig. 1.36 Wheel control actuation system: EGTS architecture

addition, the aircraft pushback procedures via tractors and tugs typically take a long time, so the use of EMAs for wheel control can conveniently speed up these phases.

From 2005 (proof of concept) to 2012 (full performance), Stirling Dynamics demonstrated the operation of an on-board electric taxiing system with the WheelTug^{TM} device [13, 57]. The system, mounted on the NLG wheels, was made of two induction motors powered by the APU. In 2011, DLR also developed a wheel control EMA for NLG wheels, which consists of two geared PMSMs installed in the wheel rim [31]. The motors, producing up to 2.25 kNm, were capable of driving an A320 at 25 km/h. Despite the relevant performances, the main drawbacks of the design were related to the use of high-performance rare-earth magnets (neodymium iron boron), which are strongly temperature dependent and not widely available. In addition, the system was developed to be powered by fuel cells, which have safety concerns related to hydrogen storage.

Other developments have been based on MLG wheels installation, which has the advantage of splitting the total power and the drawback of lower available envelope [39]. From 2011 to 2016, Safran and Honeywell Aerospace developed the Electric Green Taxiing System (EGTS), which was powered by the three-phase 115 AC bus derived from APU and included a Transformer Rectifier Unit (TRU) to supply the Wheel Actuator Control Unit with 270 V_{DC}, Fig. 1.36.

More recently, Safran (in collaboration with the University of Nottingham, Airbus, Adeneo, and DLR) also developed a direct-drive EMA to be installed at the MLG back envelope, where more space is available. The system has been proved at TRL5 level, and it includes a PMSM with extremely high torque density (42 Nm/kg), reaching an operating torque of 7000 Nm. The motor design is based on an outer rotor configuration, five-stage Halbach array magnets, double star-windings, and advanced rare-earth materials (cobalt iron and samarium cobalt).

1.4 Summary

In the first part of the chapter, the technological, environmental, societal, and market impacts of the use of electro-mechanical actuators in aircraft applications have been presented and pointed out within the contexts of the so-called "More Electric Aircraft" (MEA) concept as well as for more general trends to aircraft electrification. The MEA potential results in terms of weight reduction, energy optimization, reduction of fuel consumption, eco-compatibility, simplified integration, and operative/maintenance costs reduction have been addressed in qualitative terms (e.g., market barriers, societal gains) and quantitative terms (CO_2-emissions targets, fuel burn reduction), by highlighting open points and possible criticalities.

In the second part of the chapter, the main results deriving from the research and development programs toward the MEA objective are presented, by describing the technological path followed in the design of electrically powered actuators for aircraft applications, which started from the first EHA-VD solutions in the late 1980s, up to the flightworthy EMA solutions of recent years. The basic technological issues related to an aircraft EMA development are pointed out, by focusing the attention on each key component of the EMA (electric motor, PWM motor drive, control electronics, mechanical transmission, and fail-safe devices).

Finally, the specific concerns related to relevant actuation functions in aerospace applications are addressed, by offering a brief state-of-the-art survey on the use of EMAs for conventional functions (flight controls, landing gears, nose-wheel steering, brakes, and thrust vectoring controls), as well as for innovative functions, such as winglet movables and wheel controls.

References

1. Clean sky european research programmes, funded by the EU's FP7 and HORIZON 2020 programmes. http://www.cleansky.eu/
2. International Air Transport Association (IATA), Airline industry economic performance, end-year report, December 2019. https://www.iata.org/en/iata-repository/publications/economic-reports/airline-industry-economic-performance---december-2019---report/
3. International Air Transport Association (IATA), Airline industry economic performance, mid-year report, June 2020. https://www.iata.org/en/iata-repository/publications/economic-reports/airline-industry-economic-performance-june-2020-report/
4. International Air Transport Association (IATA), Outlook for air travel in the next 5 years. https://www.iata.org/en/iata-repository/publications/economic-reports/covid-19-outlook-for-air-travel-in-the-next-5-years/
5. International Air Transport Association (IATA), What can we learn from past pandemic episodes? https://www.iata.org/en/iata-repository/publications/economic-reports/what-can-we-learn-from-past-pandemic-episodes/
6. MOET project-Final report summary. https://cordis.europa.eu/project/rcn/81472/factsheet/en
7. More Open Electrical Technologies (MOET), EC-funded FP6 programme. https://trimis.ec.europa.eu/project/more-open-electrical-technologies
8. Single european sky initiative, SESAR joint undertaking, funded EU's HORIZON 2020 programme. https://www.sesarju.eu/

9. World health organization, coronavirus disease pandemic. https://www.who.int/emergencies/diseases/novel-coronavirus-2019
10. Ready for the more-electric aircraft (2012). https://www.safran-group.com/sites/group/files/safran_dos_presse_flipbook_gb.pdf
11. Flight path 2050 report-European vision for aviation (2017). https://ec.europa.eu/transport/sites/transport/files/modes/air/doc/flightpath2050.pdf
12. Annaz FY (2005) Fundamental design concepts in multi-lane smart electromechanical actuators. Smart Mater Struct 14(6):1227–1238. https://doi.org/10.1088/0964-1726/14/6/016
13. Aviaton Pros: WheelTug successfully tests electric drive sSystem on 737-700 (2017). https://www.aviationpros.com/aircraft/commercial-airline/news/10734494/wheeltug-tests-electric-system-that-moves-aircraft-at-less-expense-to-environment-and-bottom-line
14. Balaban E, Bansal P, Stoelting P, Saxena A, Goebel KF, Curran S (2009) A diagnostic approach for electro-mechanical actuators in aerospace systems. In: 2009 IEEE aerospace conference, pp 1–13. https://doi.org/10.1109/AERO.2009.4839661
15. Bennett JW, Atkinson GJ, Mecrow BC, Atkinson DJ (2012) Fault-tolerant design considerations and control strategies for aerospace drives. IEEE Trans Ind Electron 59(5):2049–2058. https://doi.org/10.1109/TIE.2011.2159356
16. Bennett JW, Mecrow BC, Atkinson DJ, Atkinson GJ (2011) Safety-critical design of electromechanical actuation systems in commercial aircraft. IET Electric Power Appl 5(1):37–47. https://doi.org/10.1049/iet-epa.2009.0304
17. Binder A (2012) Elektrische Maschinen und Antriebe. Springer
18. Bolam RC, Vagapov Y, Anuchin A (2018) Review of electrically powered propulsion for aircraft. In: 2018 53rd international universities power engineering conference (UPEC), pp 1–6. https://doi.org/10.1109/UPEC.2018.8541945
19. Budinger M, Liscouët J, Orieux S, Maré JC (2008) Automated preliminary sizing of electromechanical actuator architectures. Variations 3(2)
20. Budinger M, Reysset A, Halabi TE, Vasiliu C, Maré JC (2014) Optimal preliminary design of electromechanical actuators. Proc Inst Mech Engi Part G J Aerosp Eng 228(9):1598–1616. https://doi.org/10.1177/0954410013497171
21. Cao W, Mecrow BC, Atkinson GJ, Bennett JW, Atkinson DJ (2012) Overview of electric motor technologies used for more electric aircraft (mea). IEEE Trans Ind Electron 59(9):3523–3531. https://doi.org/10.1109/TIE.2011.2165453
22. Davis MA (1984) High performance electromechanical servoactuation using brushless dc motors. Technical bulletin 150
23. Derrien JC, Sécurité SD (2012) Electromechanical actuator (EMA) advanced technologies for flight controls. In: International congress of the aeronautical sciences, pp 1–10
24. Descamps D. Alexandre P, Telteu-Nedelcu D (2012) Hi-reliability electromechanical thrust vector actuation systems for European unmanned launch vehicles-a challenge for the next generation. In: Proceedings of the 5th international conference on recent advances in aerospace actuation systems and components, Toulouse (France), pp 11–158
25. Di Rito G, Luciano B, Borgarelli N, Nardeschi M (2020) Health-monitoring of a jamming-tolerant electro-mechanical actuator with differential ball screws. In: Proceedings of the 8th IEEE international workshop on metrology for aerospace, virtual/online conference, vol 8, pp 84–89. https://doi.org/10.1109/MetroAeroSpace48742.2020.9160119
26. Di Rito G, Galatolo R, Schettini F (2016) Experimental and simulation study of the dynamics of an electro-mechanical landing gear actuator. In: 30th congress of the international council of the aeronautical sciences (ICAS), Daejeon, South Korea
27. Dimino I, Gallorini F, Palmieri M, Pispola G (2019) Electromechanical actuation for morphing winglets. In: Actuators, vol 8, p 42. Multidisciplinary Digital Publishing Institute. https://doi.org/10.3390/act8020042
28. Dzikus N, Fuchte J, Lau A, Gollnick V (2011) Potential for fuel reduction through electric taxiing. In: 11th AIAA aviation technology, integration, and operations (ATIO) conference, including the AIAA balloon systems conference and 19th AIAA lighter-than, p 6931. https://doi.org/10.2514/6.2011-6931

29. Emadi K, Ehsani M (2000) Aircraft power systems: technology, state of the art, and future trends. IEEE Aerosp Electron Syst Mag 15(1):28–32. https://doi.org/10.1109/62.821660
30. Fulmer C (1996) 40 HP electro-mechanical actuator
31. Galea M, Xu Z, Tighe C, Hamiti T, Gerada C, Pickering S (2014) Development of an aircraft wheel actuator for green taxiing. In: 2014 international conference on electrical machines (ICEM), pp 2492–2498. https://doi.org/10.1109/ICELMACH.2014.6960537
32. Grand S, Valembois J (2004) Electromechanical actuators design for thrust vector control. In: Proceedings of the 2nd international conference on recent advances in aerospace actuation systems and components, Toulouse (France), pp 21–27
33. Henke M, Narjes G, Hoffmann J, Wohlers C, Urbanek S, Heister C, Steinbrink J, Canders WR, Ponick B (2018) Challenges and opportunities of very light high-performance electric drives for aviation. Energies 11(2):344. https://doi.org/doi.org/10.3390/en11020344
34. Hollinger P, Woodhouse A (2020) Airbus signals further production cut with job losses to follow. Financial Times April 29
35. Howse M (2003) All electric aircraft. Power Eng 17(4):35–37. https://doi.org/10.1049/pe:20030410
36. Iacus S, Natale F, Santamaria C, Spyratos S, Vespe M (2020) Estimating and projecting air passenger traffic during the COVID-19 coronavirus outbreak and its socio-economic impact. Saf Sci 129:104791. https://doi.org/10.1016/j.ssci.2020.104791
37. Ismail MA, Windelberg J (2017) Fault detection of bearing defects for ballscrew based electro-mechanical actuators. In: First world congress on condition monitoring (WCCM 2017). https://doi.org/10.1784/204764218823029048
38. Ismail MAA, Balaban E, Spangenberg H (2016) Fault detection and classification for flight control electromechanical actuators. In: 2016 IEEE aerospace conference, pp 1–10. https://doi.org/10.1109/AERO.2016.7500784
39. Lukic M, Hebala A, Giangrande P, Klumpner C, Nuzzo S, Chen G, Gerada C, Eastwick C, Galea M (2018) State of the art of electric taxiing systems. In: 2018 IEEE international conference on electrical systems for aircraft, railway, ship propulsion and road vehicles international transportation electrification conference (ESARS-ITEC), pp 1–6. https://doi.org/10.1109/ESARS-ITEC.2018.8607786
40. Madonna V, Giangrande P, Galea M (2018) Electrical power generation in aircraft: review, challenges, and opportunities. IEEE Trans Transp Electrif 4(3):646–659. https://doi.org/10.1109/TTE.2018.2834142
41. Manohar GA, Vasu V, Srikanth K (2018) Development of a high redundancy actuator with direct driven linear electromechanical actuators for fault-tolerance. Proc Comput Sci 133:932–939. https://doi.org/10.1016/j.procs.2018.07.089
42. Maré JC (2017) Aerospace actuators 2: signal-by-wire and power-by-wire. Wiley
43. Maré JC, Fu J (2017) Review on signal-by-wire and power-by-wire actuation for more electric aircraft. Chin J Aeronaut 30(3):857–870. https://doi.org/10.1016/j.cja.2017.03.013
44. Mazzoleni M, Maccarana Y, Previdi F, Pispola G, Nardi M, Perni F, Toro S (2017) Development of a reliable electro-mechanical actuator for primary control surfaces in small aircrafts. In: 2017 IEEE international conference on advanced intelligent mechatronics (AIM), pp 1142–1147. https://doi.org/10.1109/AIM.2017.8014172
45. Mazzoleni M, Previdi F, Scandella M, Pispola G (2019) Experimental development of a health monitoring method for electro-mechanical actuators of flight control primary surfaces in more electric aircrafts. IEEE Access 7:153,618–153,634. https://doi.org/10.1109/ACCESS.2019.2948781
46. MOOG: Aircraft capability brochure (2011). https://www.moog.com/content/dam/moog/literature/ICD/Moog-Industrial-Capabilities-Overview-en.pdf
47. Previdi F, Maccarana Y, Mazzoleni M, Scandella M, Pispola G, Porzi N (2018) Development and experimental testing of a health monitoring system of electro-mechanical actuators for small airplanes. In: 2018 26th mediterranean conference on control and automation (MED), pp 673–678 (2018). https://doi.org/10.1109/MED.2018.8442734

48. Qiao G, Liu G, Shi Z, Wang Y, Ma S, Lim TC (2018) A review of electromechanical actuators for more/all electric aircraft systems. Proc Inst Mech Eng Part C J Mech Eng Sci 232(22):4128–4151. https://doi.org/10.1177/0954406217749869
49. Rapaccini M, Saccani N, Kowalkowski C, Paiola M, Adrodegari F (2020) Navigating disruptive crises through service-led growth: The impact of COVID-19 on italian manufacturing firms. Ind Market Manag 88:225–237. https://doi.org/10.1016/j.indmarman.2020.05.017
50. Roboam X, Sareni B, Andrade AD (2012) More electricity in the air: toward optimized electrical networks embedded in more-electrical aircraft. IEEE Ind Electron Mag 6(4):6–17. https://doi.org/10.1109/MIE.2012.2221355
51. Roland Berger Ltd (2017) Aircraft electrical propulsion—onwards and upwards. https://www.rolandberger.com/publications/publication_pdf/roland_berger_aircraft_electrical_propulsion_2.pdf
52. Roland Berger Ltd. (2017) Aircraft electrical propulsion—the next chapter of aviation? https://www.rolandberger.com/publications/publication_pdf/roland_berger_aircraft_electrical_propulsion.pdf
53. Rottach M, Gerada C, Wheeler PW (2014) Design optimisation of a fault-tolerant pm motor drive for an aerospace actuation application. In: 7th IET international conference on power electronics, machines and drives (PEMD 2014), pp 1–6. https://doi.org/10.1049/cp.2014.0484
54. Schafer AW, Barrett SRH, Doyme K, Dray LM, Gnadt AR, Self R, O'Sullivan A, Synodinos AP, Torija AJ (2019) Technological, economic and environmental prospects of all-electric aircraft. Nat Energy 4(2):160–166. https://doi.org/10.1038/s41560-018-0294-x
55. Schettini F, Denti E, Di Rito G (2017) Development of a simulation platform of all-electric aircraft on-board systems for energy management studies. Aeronaut J 121(1239):710–719. https://doi.org/10.1017/aer.2017.16
56. Spitzer CR (1984) The all-electric aircraft: a systems view and proposed NASA research programs. IEEE Trans Aerosp Electron Syst AES-20(3):261–266. https://doi.org/10.1109/TAES.1984.310509
57. Stirling Dynamics (2017) Stirling redesigns the wheel! https://www.stirling-dynamics.com/news/stirling-redesigns-wheel/
58. Van Den Bossche D (2006) The A380 flight control electrohydrostatic actuators, achievements and lessons learnt. In: 25th international congress of the aeronautical sciences, pp 1–8
59. Yu ZY, Niu T, Dong HL (2018) A jam-tolerant electromechanical system. In: ACTUATOR 2018; 16th international conference on new actuators, pp 1–4

Chapter 2
Reliability and Safety of Electro-Mechanical Actuators for Aircraft Applications

Outline of the Chapter. This chapter presents the basic reliability and safety concerns related to the airworthiness certification of airborne systems, with a special focus on Electro-Mechanical Actuators (EMAs).

Section 2.1, starting from the definition of fault distribution models for industrial components, presents the reliability and safety requirements for the airworthiness certification of airborne systems, and points out the necessity of using architectures with hardware and/or analytical redundancies.

Section 2.2 provides a survey on the application of redundancy concepts on EMAs, with relevant examples of fault-tolerant designs of power electronics, electric motors, and mechanical transmissions.

Section 2.3 describes the standard guidelines, methods, and procedures to be applied for the System Safety Assessment of complex airborne systems, with special focus on the Functional Hazard Assessment (FHA), the Fault-Tree Analysis (FTA), and the Failure Modes and Effects and Criticality Analysis (FMECA), which are presented by also using practical examples. A categorization of system Built-in Tests (BITs) is then given, by differentiating them in terms of monitoring objectives.

Section 2.4 is finally devoted to a practical example of preliminary system safety assessment, carried out on the EMA system of the morphing flaps of a more electric aircraft demonstrator.

2.1 Basic Reliability and Safety Concerns

This section reviews fundamental reliability and safety issues from certification requirements in the aerospace sector to the concepts of hardware and analytical redundancy.

© Springer Nature Switzerland AG 2021

M. Mazzoleni et al., *Electro-Mechanical Actuators for the More Electric Aircraft*, Advances in Industrial Control, https://doi.org/10.1007/978-3-030-61799-8_2

2.1.1 Fault Regimes of Airborne Components

All industrial components are typically prone to three types of fault regimes [19]:

- *Burn-in* faults, which are related to design errors or materials' imperfections and occur in the initial phases of components' life;
- *Random* faults, which are related to nondeterministic factors (e.g., overloads) and occur throughout the components' life;
- *Wear-out* faults, which are related to materials' aging and occur in the final phases of components' life.

By supposing that the component lifetime t is continuous (the description in discrete-time terms is very similar), each fault regime can be characterized by a Probability Density Function (PDF) $f(t)$, i.e., *failure density function* that represents how the increment of fault probability is distributed along the component life. From there, other quantities can be derived:

- *failure probability* $F(t)$, i.e., the probability that, at time t, the fault is occurred;
- *survivability probability (or reliability)* $S(t) = 1 - F(t)$, i.e., the probability that, at time t, the fault is not occurred;
- *hazard (or failure) rate* $h(t)$, i.e., the increment of fault probability referred to the components survived at time t.

The failure probability $F(t)$ and the hazard rate $h(t)$ are related to $f(t)$ via (2.1), where t is the component lifetime:

$$F(t) = \int_0^t f(t)\,dt, \tag{2.1a}$$

$$h(t) = \frac{f(t)}{1 - F(t)} = \frac{f(t)}{S(t)}. \tag{2.1b}$$

Another useful quantity is the Mean Time Between Failures (MTBF), which can be thought as the expected time between two failures for a repairable system:

$$\text{MTBF} = \int_0^\infty S(t)\,dt = \int_0^\infty t \cdot f(t)\,dt. \tag{2.2}$$

The burn-in, random, and wear-out fault regimes can be effectively represented by the following PDFs (though other models are also possible):

$$f_b\left(t|\beta_b, \eta_b\right) = \frac{\beta_b}{\eta_b}\left(\frac{t}{\eta_b}\right)^{\beta_b - 1} \cdot \exp\left[-\left(\frac{t}{\eta_b}\right)^{\beta_b}\right], \tag{2.3a}$$

$$f_w\left(t|t_w, \sigma_w\right) = \frac{1}{\sigma_w\sqrt{2\pi}} \cdot \exp\left[-\frac{1}{2}\left(\frac{t - t_w}{\sigma_w}\right)^2\right], \tag{2.3b}$$

$$f_r\left(t|\lambda_r\right) = \lambda_r \cdot \exp\left[-\lambda_r t\right], \tag{2.3c}$$

where β_b and η_b are the shape and scale factors of the Weibull distribution related to burn-in faults, λ_r is the failure rate of the the exponential distribution related to random faults, while t_w and σ_w are the central value and the standard deviation of the Gaussian distribution related to wear-out faults.

The shape parameter β_b of the Weibull distribution characterizes the aging property of the components. In particular, [16]:

- if $\beta_b < 1$ the failure rate decreases over time (negative aging);
- if $\beta_b = 1$ the failure rate is constant over time (non-aging), and the distribution becomes exponential;
- if $\beta_b > 1$ the failure rate increases with time (positive aging), which is appropriate for modeling wear-out failure due to gradual deterioration/degradation of an item over time.

In the case of a Weibull distribution, the following expressions hold:

$$F_b\left(t|\beta_b, \eta_b\right) = 1 - \exp\left[-\left(\frac{t}{\eta_b}\right)^{\beta_b}\right], \tag{2.4a}$$

$$h_b\left(t|\beta_b, \eta_b\right) = \frac{f_b(t)}{1 - F_b(t)} = \frac{\beta_b}{\eta_b}\left(\frac{t}{\eta_b}\right)^{\beta_b - 1}, \tag{2.4b}$$

$$S_b\left(t|\beta_b, \eta_b\right) = 1 - F_b(t) = \exp\left[-\left(\frac{t}{\eta_b}\right)^{\beta_b}\right]. \tag{2.4c}$$

To provide a practical interpretation of fault regimes' in the lifetime of industrial components, once defined three weighing parameters k_b, k_r, and k_w with $0 \leq k_b \leq 1$, $0 \leq k_r \leq 1$, and $0 \leq k_w \leq 1$, the total PDF $f_T(t|\boldsymbol{\vartheta})$ and hazard rate $h_T(t|\boldsymbol{\vartheta})$, characterizing the faults of a statistical sample of industrial components, can be expressed by

$$f_T(t|\boldsymbol{\vartheta}) = k_b \cdot f_b(t|\boldsymbol{\vartheta}) + k_r \cdot f_r(t|\boldsymbol{\vartheta}) + k_w \cdot f_w(t|\boldsymbol{\vartheta}), \tag{2.5a}$$

$$F_T(t|\boldsymbol{\vartheta}) = \int_0^t f_T(t|\boldsymbol{\vartheta})\, dt, \tag{2.5b}$$

$$h_T(t|\boldsymbol{\vartheta}) = \frac{f_T(t|\boldsymbol{\vartheta})}{1 - F_T(t|\boldsymbol{\vartheta})} = \frac{f_T(t|\boldsymbol{\vartheta})}{S_T(t|\boldsymbol{\vartheta})}, \tag{2.5c}$$

where $\boldsymbol{\vartheta}$ represents the union of the parameters of the distributions $f_b(t)$, $f_r(t)$, $f_w(t)$, and $k_b + k_r + k_w = 1$.

It is worth noting that, typically, $k_r \approx 1$: the impact of burn-in faults is actually minimized via design process optimization, while the good predictability of wear-out faults permits to maintain them outside the operation lifetime via maintenance programs.

Typical qualitative results in terms of PDF and failure rate as function of component life are reported in Fig. 2.1, obtained by assuming $\beta_b = 0.5$, $\eta_b = 100/\lambda_r$,

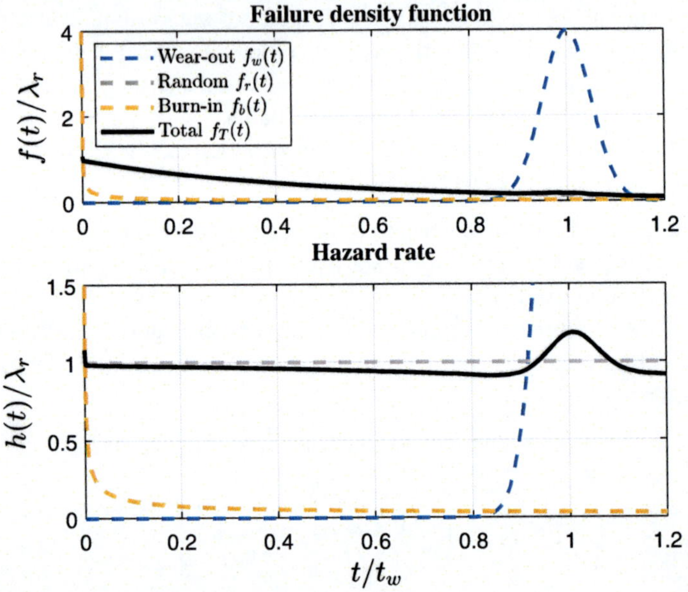

Fig. 2.1 Failure rates and PDF for different fault regimes

$\sigma_w = 0.1/\lambda_r, t_w = 2/\lambda_r, k_r = 0.98$ and $k_b = k_w = 0.01$ in (2.3)–(2.5). It can be noted that the total fault regime is mainly dominated by the random faults [19]. Notice that the hazard rate represents an instant failure probability that varies with lifetime. Only in specific cases (e.g., when $f(t)$ is modeled as an exponential PDF) we obtain a constant failure rate s.t. $h(t) = \lambda$. In this special case, we have that MTBF $= \frac{1}{\lambda}$.

Though the fault regimes for airborne components clearly depend on environment and loads [1, 13, 15, 18, 19, 23], it is typically observed that random failure rates range:

- between 10^{-5} and 10^{-7} h^{-1} for electronic components;
- between 10^{-4} and 10^{-6} h^{-1} for electrical components;
- between 10^{-8} and 10^{-9} h^{-1} for mechanical components.

2.1.2 Airworthiness Certification Requirements

Reliability and safety are primary factors for enabling any technology to be called "flightworthy". The airworthiness certification clearly depends on the vehicle type (aircraft, helicopters, UAS), application (civil, military), and category (large or small size, propulsion type) [7, 8, 19, 24], but all certification procedures aim to verify that the examined vehicle is capable of satisfying specific requirements in terms of performances and reliability/safety levels.

For this reason, reliability/safety activities have to be developed and integrated into the whole design process of on-board systems, by means of the so-called Reliability Availability Maintainability and Safety (RAMS) discipline.

The airworthiness certifications standards related to large aircraft and rotorcraft for civil applications [7, 8] agree in classifying the failure conditions related to loss of functions in terms of severity effects, see Table 2.1:

- *Catastrophic* failures are related to the vehicle loss or to fatalities of occupants/crew;
- *Hazardous* failures are related to strong reductions of vehicle performances/safety margins or to serious/fatal injuries of occupants/crew;
- *Major* failures are related to the mission loss, or to significant injuries/discomfort of occupants/crew, or significant increase of crew workload;
- *Minor* failures are related to slight reductions of vehicle performances/safety margins, or slight discomfort of occupants, or a slight increase of crew workload;
- *No safety effect* failures are related to a negligible reduction of vehicle performances or slight discomfort of occupants (no effects on flight crew).

The certifications standards also indicate a relationship between the severity effects and the allowable probability of occurrence of the failure condition, Fig. 2.2:

- *Catastrophic* failures are allowed to be *extremely improbable*, i.e., their probability occurrence must be lower than 10^{-9} per flight hour;
- *Hazardous* failures are allowed to be *extremely remote*, i.e., their probability occurrence must be lower than 10^{-7} per flight hour;
- *Major* failures are allowed to be *remote*, i.e., their probability occurrence must be lower than 10^{-5} per flight hour;
- *Minor* failures are allowed to be *probable*, i.e., their probability occurrence must be lower than 10^{-3} per flight hour;
- *No safety effect* failures are not strictly related to probability requirements, but they are defined with reference to maintenance issues.

In addition, when the failure condition involves software-related functions, the requirements are expressed in terms of Development Assurance Level (DAL) [25], and the relationship between the severity effects and the allowable DAL is given in Table 2.2, where the DAL increases from E to A depending on the required SW test coverage. In particular, there are three types of coverage:

- *Statement coverage*, which verifies that every statement in the SW is invoked at least once (due to demonstrate the DAL C);
- *Decision coverage*, which verifies that every point of entry and exit in the SW is invoked at least once and every decision is taken all possible outcomes at least once (due to demonstrate the DAL B);
- *Modified condition/decision coverage*, which verifies that every point of entry and exit in the SW is invoked at least once, every condition in a decision is taken all possible outcomes at least once, every decision is taken all possible outcomes at least once, and each condition in a decision is shown to independently affect the decision's outcome (due to the demonstration of the DAL A).

Table 2.1 Classification of failure conditions

Classification of failure conditions		No safety effect	Minor	Major	Hazardous	Catastrophic
Severity of the effect	Effect on airplane	No effect of operational capabilities or safety	Slight reduction in functional capabilities or safety margin	Significant reduction in functional capabilities or safety margin	Large reduction in functional capabilities or safety margin	Normally with hull loss
	Effect on occupants excluding flight crew	Inconvenience	Physical discomfort	Physical distress, possibly including injuries	Serious or fatal injury to a small number of passengers of cabin crew	Multiple fatalities
	Effect on flight crew	No effect	Slight increase in workload	Physical discomfort or a significant increase in workload	Physical distress or excessive workload impairs ability to perform tasks	Fatalities or incapacitation

If the typical failure rates characterizing airborne components (Sect. 2.1.1) are compared with the airworthiness certification requirements, it is clear that they are far from adequate in terms of reliability levels. For this reason, on-board systems often apply redundancy, which can be classified into two main categories:

- *Hardware redundancy*, when redundant components are physically integrated into the equipment;
- *Analytical redundancy*, when the components' functions are simulated via mathematical models into the control software to monitor and validate the outputs of the physical components.

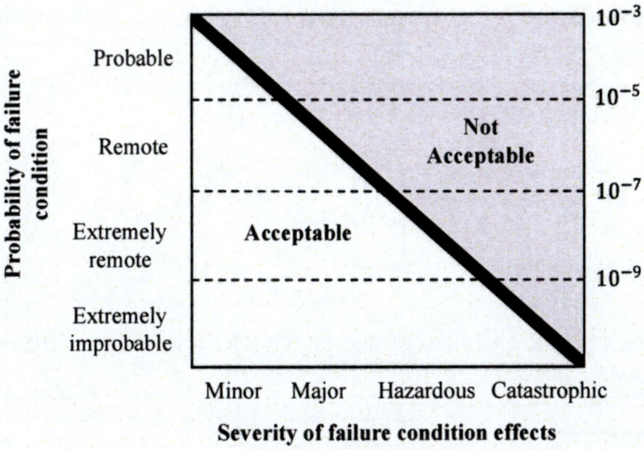

Fig. 2.2 Relationship between the severity effects and the allowable probability of occurrence of the failure condition

Table 2.2 Required development assurance level for airborne SW

Failure condition severity	SW test coverage			DAL
	Modified condition/decision coverage	Decision coverage	Statement coverage	
Catastrophic	Yes	Yes	Yes	A
Hazardous	No	Yes	Yes	B
Major	No	No	Yes	C
Minor	No	No	No	D
No safety effect	No	No	No	E

2.1.3 *Hardware Redundancy*

With particular reference to airborne EMAs (similar concepts are applicable to any on-board system), hardware redundancy can be applied at different levels, see Fig. 2.3:

- *Load-level redundancy* implies that a specific actuation function (e.g., flight control, landing gear, brake, etc.) is split into redundant load paths, each one driven by a single simplex EMA, Fig. 2.3 (left);
- *Actuator-level redundancy* implies that a specific actuation function is obtained by a single load path, driven by redundant simplex EMAs, Figs. 2.3 (middle) and 2.4;
- *Subsystem-level redundancy* implies that a specific actuation function is obtained by a single load path, driven by a single fault-tolerant EMA integrating redundant components, Figs. 2.3 (right) and 2.5.

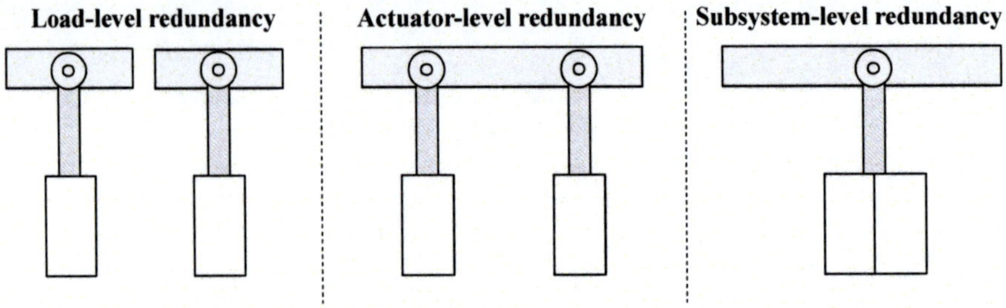

Fig. 2.3 Redundancy concepts: (left) load level; (middle) actuator level; (right) subsystem level

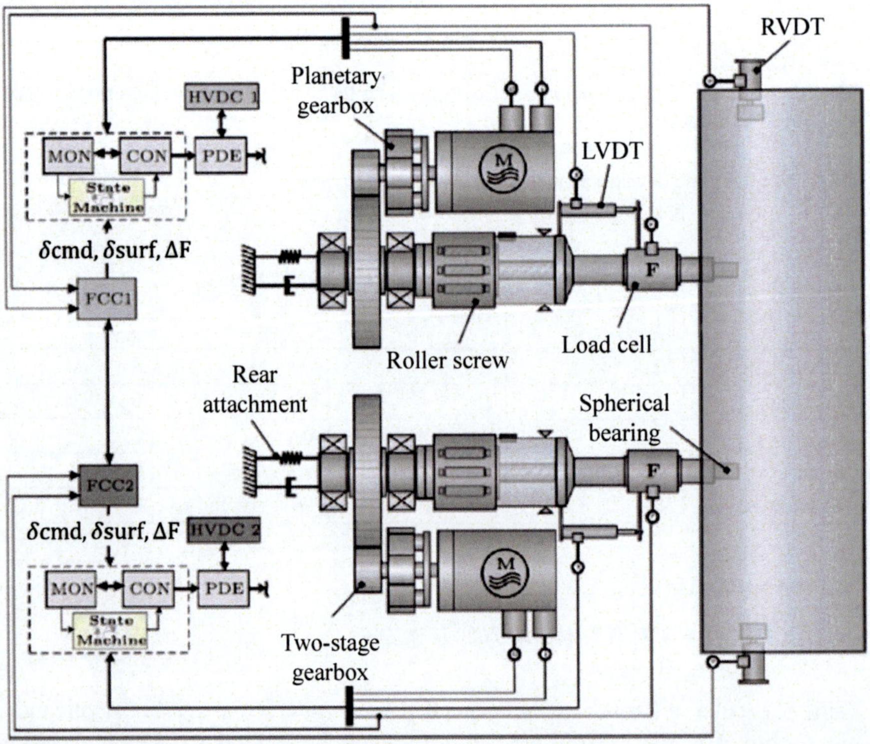

Fig. 2.4 Actuator-level redundancy: dual EMAs in force-summing arrangement. Reprinted from [4]: Model-based design and experimental verification of a monitoring concept for an active-active electromechanical aileron actuation system, 94, Arriola, D., Thielecke, F., 322–345, Copyright (2017), with permission from Elsevier

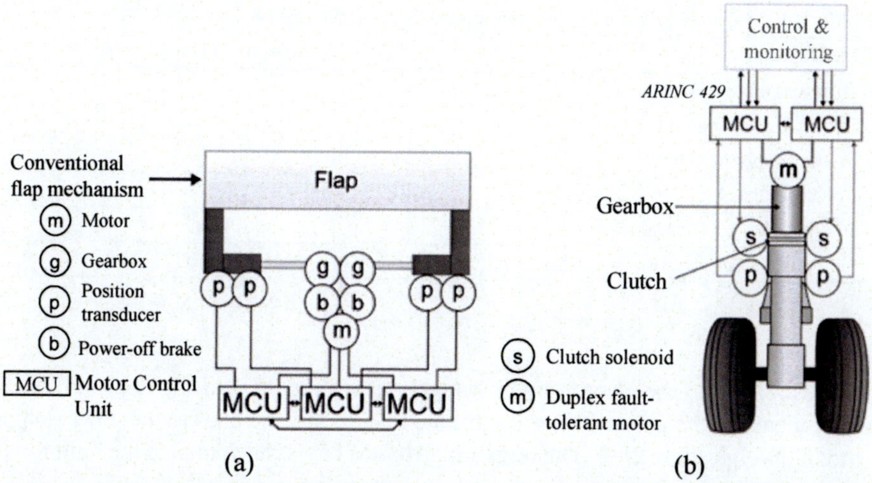

Fig. 2.5 Subsystem-level redundancy: **a** EMA with triplex ECU; **b** EMA with dual ECU. Republished with permission of Institution of Engineering and Technology (IET), from [5]: Safety-critical design of electromechanical actuation systems in commercial aircraft, Bennett, J.W., Mecrow, B.C., Atkinson, D.J., Atkinson, G.J., 5(1) © (2003); permission conveyed through Copyright Clearance Center, Inc.

2.1.4 Analytical Redundancy

As previously outlined, redundancy can be applied by physically installing additional components devoted to the same function (*hardware redundancy*) or by simulating the function via mathematical models, in order to consolidate the hardware outputs (*analytical redundancy*). In some advanced sensing applications, the mathematical simulations can be used as unique source of information, thus obtaining the so-called *virtual sensors* [11, 14, 20, 21, 28–30].

Analytical redundancy is thus a form of real-time model-based monitoring, since the function models must be executable in real-time by the equipment SW, in order to permit the model outputs to be synchronized and available together with the ones derived from hardware components (Sect. 4.3).

2.2 Fault-Tolerant Electro-Mechanical Actuator Solutions

The severe requirements in terms of compactness, electrical stress, environmental conditions, and vibration levels assigned to the electric motors and to the power/control electronics of airborne EMAs imply that simplex solutions are often not suitable for airworthiness. As a reference guideline, the major fault modes contributing to the Mean Time Between Failures (MTBF) of the electric/electronic section of a simplex EMA are shown in Table 2.3. It can be noted that the fault

Table 2.3 Major electric/electronic fault modes for a simplex EMA

Fault mode	MTBF contribution (%)
Open circuit (single phase)	6
Short circuit (single phase)	3
Power supply module	24
Power bridge	40
I/O module	5
Computation/processor	4

modes related to the electric motor roughly cover the 25% of the total MTBF, the fault modes to the power bridge contribute to 40% and the ones to the power supply module weigh about 25%. The remaining 10% MTBF is attributed to the fault modes of the low-power electronics.

In addition, the management of the mechanical jamming fault remains an open point for EMAs, and it probably represents the most relevant barrier to the use of EMAs for safety-critical functions.

For these reasons, the implementation of fault-tolerant solutions is crucial to enhance the EMA reliability/safety, and many research efforts have been made to develop fault-tolerant architectures for power electronics, electric motors, and mechanical transmissions.

2.2.1 Fault-Tolerant Electronics

Some examples of fault-tolerant power electronics are reported in Fig. 2.6. The phase-isolating design in Fig. 2.6 (top left) permits to drive each motor phase by a full H-bridge, so that the power bridge is composed of twelve MOSFET switches. With this solution, the motor is capable of operating with minor performance degradation even if one MOSFET (or one phase) fails [5].

The triple three-phase power bridge in Fig. 2.6 (top right) is instead composed of three conventional power bridge with six MOSFETs driving a separate three-phase system. The solution clearly has the drawback of increased number of components, but it has the advantage to implement standard motor drive techniques [5].

The four-leg converter in Fig. 2.6 (bottom) finally provides a fault-tolerant solution with only eight MOSFET switches, in which the additional "leg" is in stand-by when the system normally operates, and it is activated to control the neutral point of the three-phase system when a fault occurs. The main drawback of the solution is related to the management of the failure transient [6].

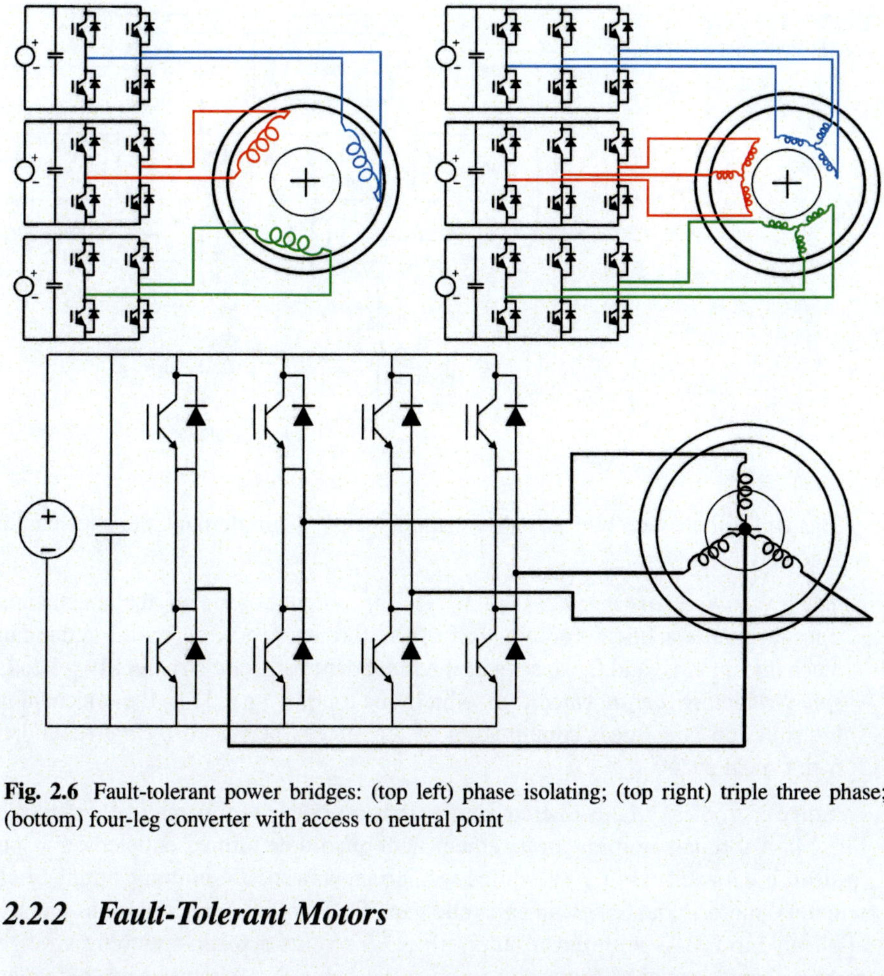

Fig. 2.6 Fault-tolerant power bridges: (top left) phase isolating; (top right) triple three phase; (bottom) four-leg converter with access to neutral point

2.2.2 *Fault-Tolerant Motors*

Many R&D activities have been carried out for the development of electric machines with redundant phases, and two relevant examples are given in Fig. 2.7, where a double three-phase system is used, so that each three-phase system is coupled with a magnets' section on a unique output shaft. In [12], a more compact solution is proposed, with a fault-tolerant PMSM with five phases.

2.2.3 *Jamming-Tolerant Mechanical Transmissions*

The counteraction of mechanical transmission jamming plays a key role in the development of aircraft EMAs. The problem is not only to disconnect the load from the jammed elements but also to maintain the operability after the fault. This is why many R&D efforts have been dedicated to the design of the fault-tolerant mechanical transmission, with reference to both rotary and linear output [17, 31].

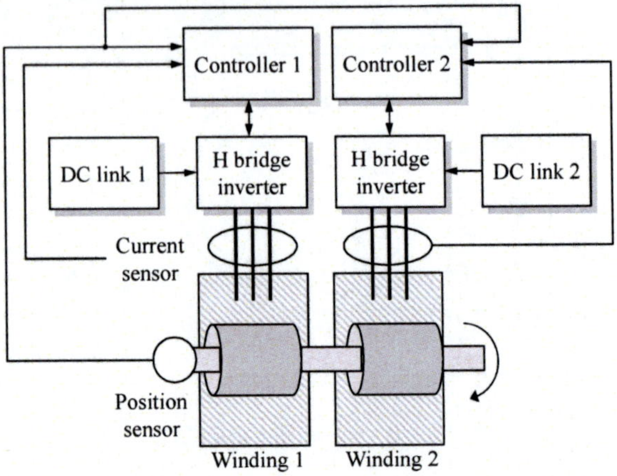

Fig. 2.7 Fault-tolerant PMSM with dual three-phase system

Basically, there are two possible arrangements to implement the mechanical redundancy [2]:

- *torque-summing arrangement*, in which the output torque of the mechanical transmission is a linear combination of the torques produced by the redundant mechanical paths, and the speeds of the redundant paths are structurally related;
- *speed-summing arrangement*, in which the output speed of the mechanical transmission is a linear combination of the speeds produced by the redundant mechanical paths.

Some examples of fault-tolerant rotary transmissions are schematically shown in Fig. 2.8. A torque-summing arrangement with quadruple motors coupled on a spur gearbox is depicted in Fig. 2.8a, while Fig. 2.8b shows a speed-summing arrangement with dual motors coupled on an epicyclic gear. Fault-tolerant transmissions are also applicable to EMAs with linear output: Fig. 2.9 reports a speed-summing arrangement with differential ball screw assemblies, while Fig. 2.10 shows a high-redundant torque-summing solution with three clusters of three motors working in parallel.

2.3 Approach to the System Safety Assessment

2.3.1 Guidelines, Methods, and Procedures

In order to harmonize and codify the methods and the procedures for the safety assessment of complex airborne systems aiming to the airworthiness certification, the following reference documents are used, Fig. 2.11:

- SAE ARP 4754 "*Certification considerations for highly-integrated or complex aircraft systems*", providing guidelines for the development and certification of aircraft systems [3];

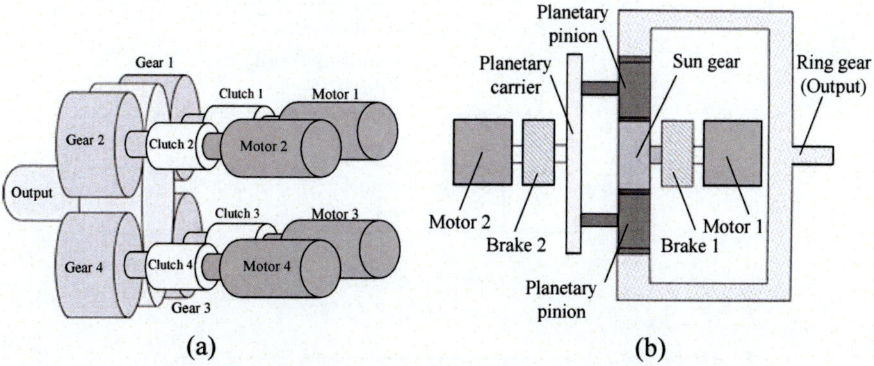

Fig. 2.8 Fault-tolerant rotary transmissions: **a** torque summing; **b** speed summing

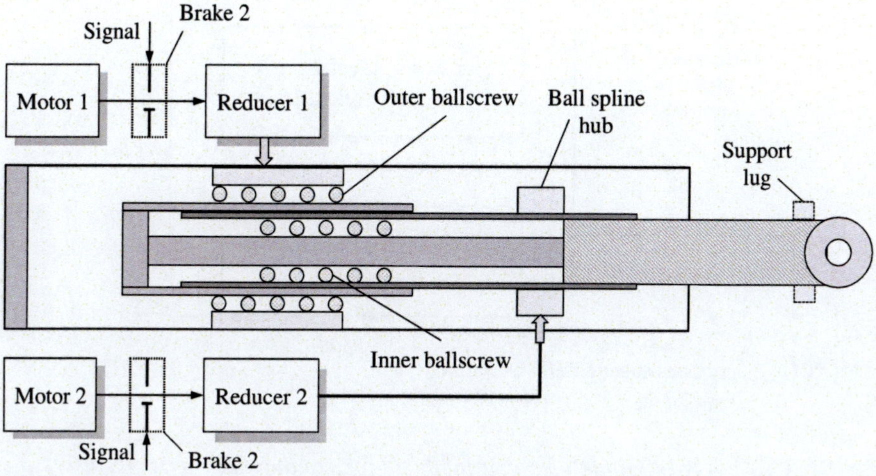

Fig. 2.9 Fault-tolerant linear transmission with nested screws

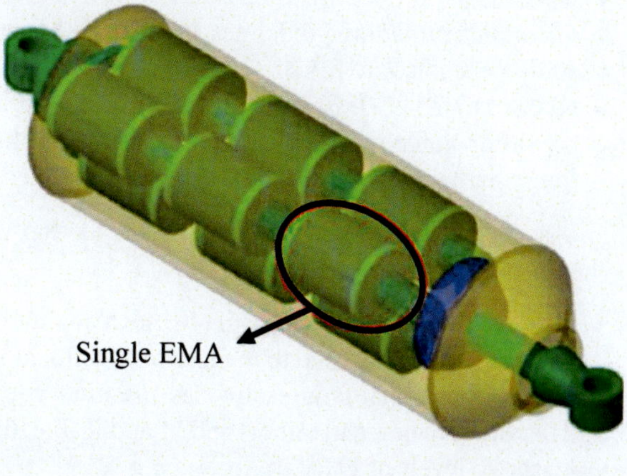

Fig. 2.10 Fault-tolerant linear transmission with 3 × 3 cluster motors. Reprinted from [17]: Development of a high redundancy actuator with direct driven linear electromechanical actuators for fault-tolerance, 133, Manohar, G.A., Vasu, V., Srikanth, K., 932–939, Copyright (2018), with permission from Elsevier

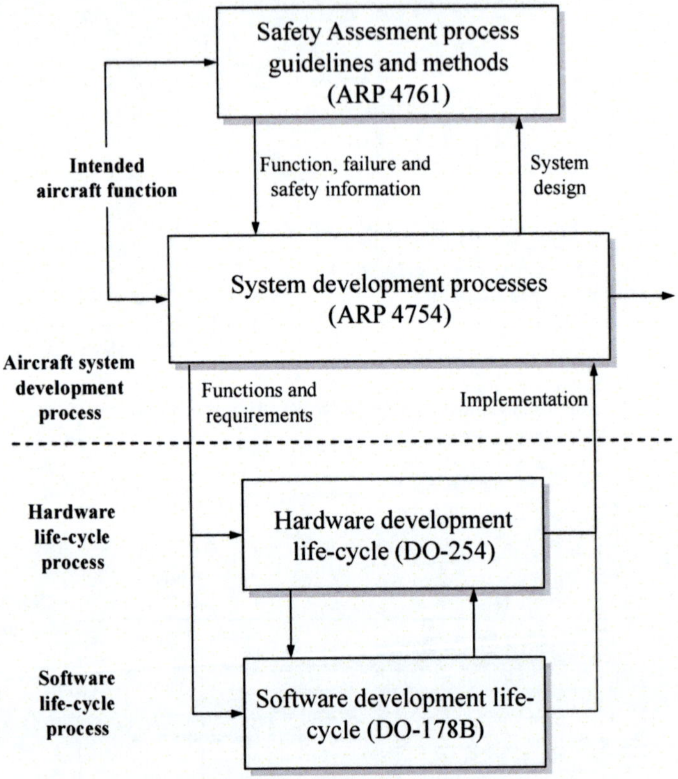

Fig. 2.11 Safety assessment general workflow

- SAE ARP 4761 "*Guidelines and methods for conducting the safety assessment process on civil airborne systems and equipment*", providing indications on the tools and techniques to be used for the safety assessment of aircraft systems [27];
- RTCA DO-178C "*Software considerations in airborne systems & equipment certification*", providing guidelines for the development and certification of airborne equipment software [25];
- RTCA DO-254 "*Design assurance guidance for airborne electronic hardware*", providing guidelines for the development and the certification of airborne electronic hardware [26].

As shown by Figs. 2.12, 2.13 and 2.14, the RAMS activities are articulated into the following steps:

- *Functional Hazard Assessment (FHA)*, a qualitative analysis aiming to define the potential hazards related to the loss of functional requirements in specific mission phases. Functional failures are classified in terms of consequences (catastrophic, hazardous, major, and minor) and, if applicable, the adequate DAL is assigned. At least two levels of FHA are foreseen:

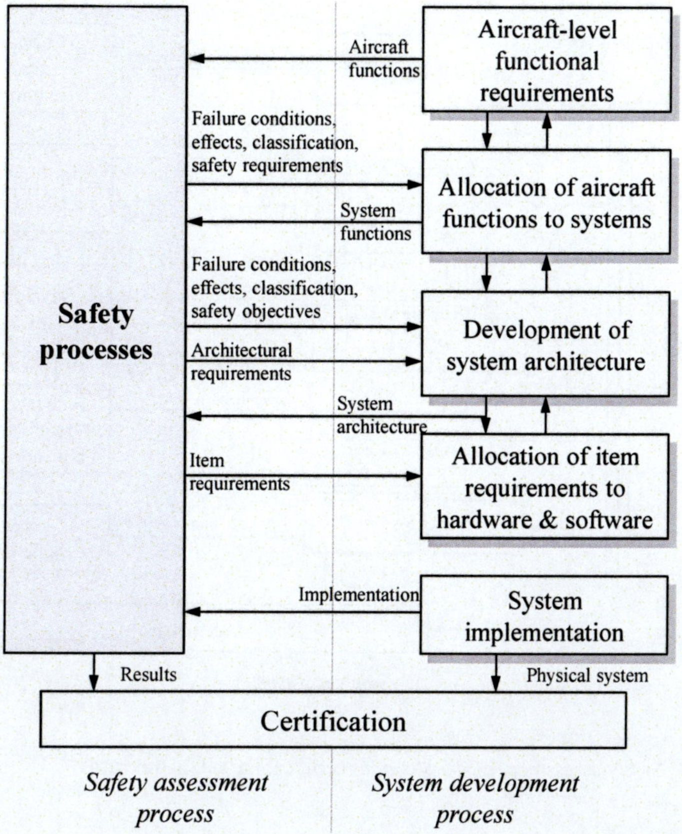

Fig. 2.12 Safety assessment process toward certification: development phases

- *vehicle-level FHA*, addressing the vehicle functions, in which the safety budgets of vehicle functional failures are allocated to systems;
- *system-level FHA*, addressing the system functions, in which the safety budgets of systems' functions are allocated to subsystems;

• *Preliminary System Safety Assessment (PSSA)*, supporting the system architecture design, in which diverse systems' architectures (differing in technologies, working concepts, and redundancies) are compared in terms of RAMS features via

- *Fault-Tree Analysis (FTA)*, a logic flowchart which defines the dependence between the faults to system parts for a specific failure case (Sect. 2.3.3);
- *Failure Mode and Effect Analysis (FMEA)*, a table in which the failure modes of each system part are classified and qualitatively analyzed in terms of effects to higher level and lower level parts (Sect. 2.3.4);
- *Reliability Block Diagrams (RBD)*, a logic diagram which defines the dependence between the reliability of system parts and the reliability of a specific system function;

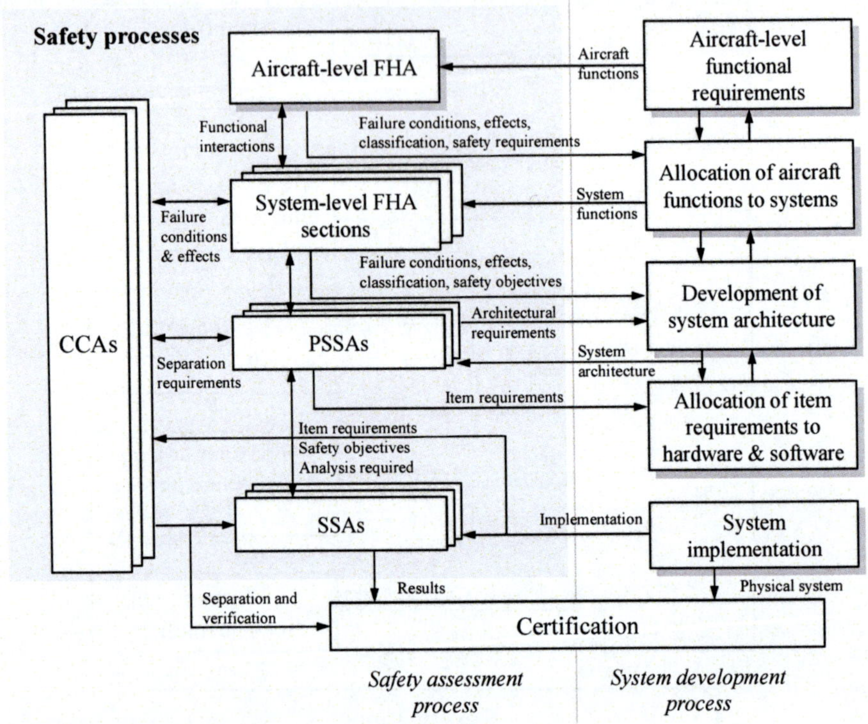

Fig. 2.13 Safety assessment process towards certification: safety processes

- The PSSA is thus a *top-down* activity that aims at allocating the reliability/safety requirements from systems to components. Its outputs are the system architecture definition, the reliability requirements of hardware parts, and the DAL of software units.

- *System Safety Assessment (SSA)*, supporting the system design and implementation, in which the developed hardware and software are analyzed/verified via FTA and
  - *Failure Mode, Effects, and Criticality Analysis (FMECA)*, a table in which the failure modes of each system part are both qualitatively analyzed in terms of effects to higher level and lower level parts and quantitatively evaluated in terms of probability of occurrence (Sect. 2.3.4);
  - *Failure Mode and Effect Summary (FMES)*, a table in which the major FMECA results are reported, by highlighting the most relevant outcomes to be addressed for airworthiness certification requirements;
  - The SSA is thus a *bottom-up* activity that aims at verifying that the system design can fulfill the safety/reliability requirements for airworthiness certification.

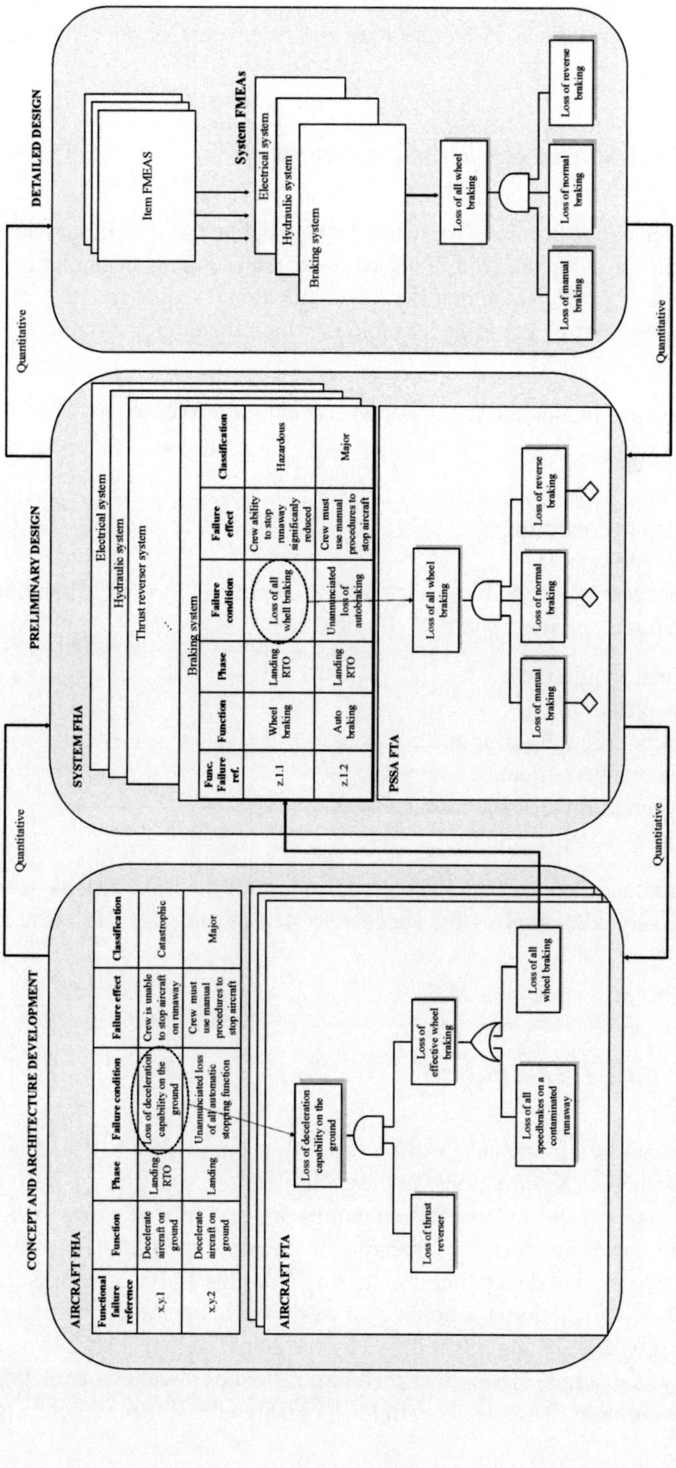

Fig. 2.14 Relationships between FHAs and FTA/FMEAs

- *Common Cause Analysis (CCA)*, which is performed throughout the safety processes to identify potential common-mode faults or single-point-of-failures, via Zonal Safety Analysis, Particular Risk Analysis, and Common-Mode Analysis.

2.3.2 Functional Hazard Assessment

The first step to be accomplished in a FHA is to define the system functional requirements, which are derived from the functional requirements of higher level systems (or the vehicle itself). Once that the list of functional requirements is obtained, the FHA is carried out by compiling a table in which the columns report the following information/outcomes:

- Failure code, an alphanumerical code containing references to
  - the system;
  - the functional requirement;
  - the failure condition;
  - the mission phase;
  - As an example, the failure code "S1.2.3.A" means "System 1, function 2, failure condition 3, mission phase A", Table 2.3.
- Functional requirement;
- Mission phase;
- Failure condition description;
- Failure condition effects;
- Failure condition classification;
- Remarks/mitigating actions.

The basic outcomes of the FHA are the Most Critical Conditions, which have to be specifically addressed by the successive step of the safety assessment: the FTA (Table 2.4).

2.3.3 Fault-Tree Analysis

The FTA can be performed in different design phases, by carrying out the logic workflow in opposite directions. In the preliminary design phases, starting from the FHA outcomes (most critical failure conditions and related safety budgets) a top-down workflow is operated, by allocating the reliability requirements from system to parts. In the detailed design phases, starting from the FMECA results (failure rates and reliability predictions), a bottom-up workflow is operated, aiming to verify that the reliability/safety budget is fulfilled by the actual design (Fig. 2.15).

A relevant example of top-down FTA aiming to the allocation of reliability requirements for airborne EMAs is given in [10], in which the PSSA activities for a MALE

Table 2.4 FHA table template

Failure code	Function	Phase	Failure condition	Failure effects	Classification	Remarks/ mitigating action
…	…	…	…	…	…	
S1.2.3.A	Description of system functional require-ment n. 2	Description of mission phase A	Failure condition n. 3 causing the loss of function 2	Description of failure effects (if possible, mentioning the effects on occupants, flight crew and vehicle perfor-mances)	Catastrophic/ Hazardous/ Major/Minor/ No safety effect	Field for notes, resulting criticalities, potential mitigations, link to other failure codes
…	…	…	…	…	…	

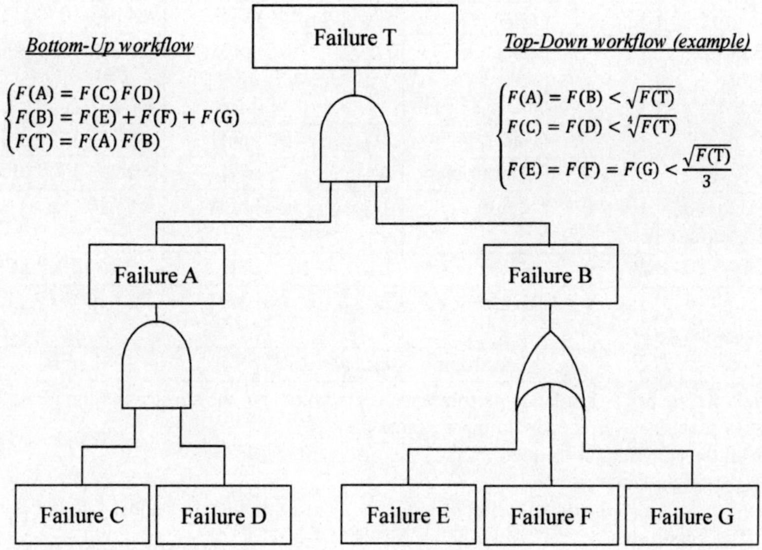

Fig. 2.15 FTA example with bottom-up and top-down workflow

Table 2.5 Failure rate budgeting for flight control EMAs of a MALE UAS

	FHA outcomes		EMA failure rate budgeting[d]	
Failure condition	Failed surfaces	Effects	Reliability constraint[c]	Derived requirement (λ_{EMA})
Partial loss[a] of aileron surface	1	Major	$4\lambda_{EMA} < 10^{-5}$ pFH	$<2.50 \cdot 10^{-6}$ pFH
	2, on different wings	Major	$4\lambda_{EMA}^2 < 10^{-5}$ pFH	$<1.58 \cdot 10^{-3}$ pFH
	2, on different wings	Hazardous	$2\lambda_{EMA}^2 < 10^{-7}$ pFH	$<2.24 \cdot 10^{-4}$ pFH
Total loss[b] of aileron surface	1	Hazardous	$4\lambda_{EMA} < 10^{-7}$ pFH	$<2.50 \cdot 10^{-8}$ pFH
	2	Catastrophic	$6\lambda_{EMA}^2 < 10^{-8}$ pFH	$<4.08 \cdot 10^{-5}$ pFH
Partial loss of elevator surface	1	Minor	$3\lambda_{EMA} < 10^{-3}$ pFH	$<3.33 \cdot 10^{-4}$ pFH
	2	Major	$3\lambda_{EMA}^2 < 10^{-5}$ pFH	$<1.83 \cdot 10^{-3}$ pFH
	3	Hazardous	$\lambda_{EMA}^3 < 10^{-7}$ pFH	$<4.64 \cdot 10^{-3}$ pFH
Total loss of elevator surface	1	Hazardous	$3\lambda_{EMA}^2 < 10^{-7}$ pFH	$<3.33 \cdot 10^{-8}$ pFH
	2	Catastrophic	$3\lambda_{EMA}^3 < 10^{-8}$ pFH	$<5.77 \cdot 10^{-5}$ pFH
	3	Catastrophic	$\lambda_{EMA}^3 < 10^{-8}$ pFH	$<2.15 \cdot 10^{-3}$ pFH
Partial loss of rudder surface	1	Major	$2\lambda_{EMA} < 10^{-5}$ pFH	$<5 \cdot 10^{-6}$ pFH
	2	Hazardous	$\lambda_{EMA}^2 < 10^{-7}$ pFH	$<3.16 \cdot 10^{-4}$ pFH
Total loss of rudder surface	1	Hazardous	$2\lambda_{EMA} < 10^{-7}$ pFH	$<5 \cdot 10^{-8}$ pFH
	2	Catastrophic	$\lambda_{EMA}^2 < 10^{-8}$ pFH	$<10^{-4}$ pFH

[a]Partial loss implies a significant performance degradation, but the surface motion is still safe
[b]Total loss causes a surface jam or unsafe motion
[c]Reliability/safety requirements from [24]
[d]pFH: per Flight Hour
Reproduced from [10] originally published open access and licensed under CC-BY 3.0. https://journals.sagepub.com/doi/pdf/10.1177/1687814016644576

UAS FCS are addressed (see Sect. 4.2.4). Starting from the FHA outcomes (the FCS architecture is reported in Fig. 2.19a) and applying the reliability/safety requirements given in [24], the EMA failure rate requirement is derived as reported in Table 2.5.

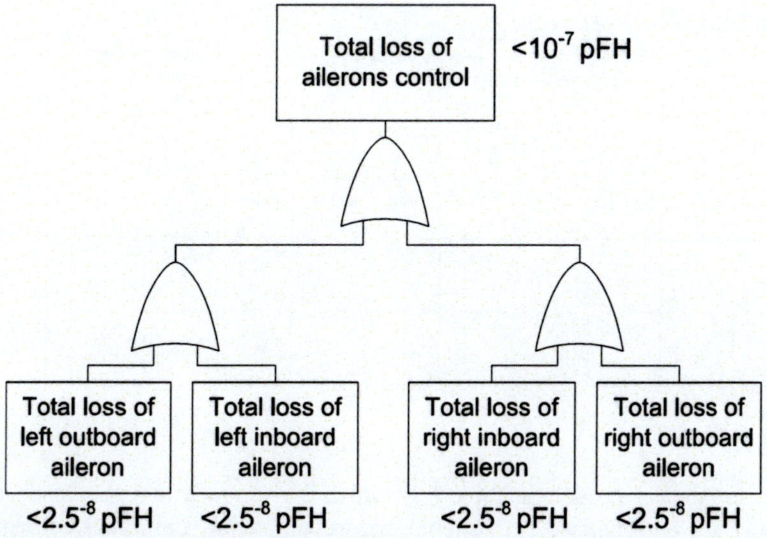

Fig. 2.16 MALE UAS FCS PSSA: EMA failure rate allocation from FHA outcomes. Reproduced from [10]—originally published open access and licensed under CC-BY 3.0. https://journals.sagepub.com/doi/pdf/10.1177/1687814016644576

The failure rate calculation (λ_{EMA} in Table 2.5) is made under the assumption that the EMA is only affected by random faults (i.e., burn-in and wear-out effects are neglected), so that its failure rate is constant [19]. The results demonstrate that the roll control function is the most critical one and that the dimensioning criterion for the EMA failure rate is related to the total loss of 1 out of 4 ailerons (Figs. 2.16 and 2.17).

2.3.4 *Failure Mode, Effects, and Criticality Analysis*

The FMECA is a reliability procedure that determines for each system fault, the modes of fault occurrence, and its effect on operations of higher level and lower level systems. The FMECA also aims at identifying single-point-of-failures and (depending on the design phase) at quantitatively evaluating the probability of occurrence of the faults. The FMECA is the result of two reliability procedures: the FMEA, which provides qualitative indications, and the Criticality Analysis (CA), which also includes quantitative estimations of fault probability.

The FMEA, the CA, and the FMECA use a bottom-up approach and they are carried out by compiling specific tables (an FMEA template is given in Table 2.6). Starting from the lowest level of the system hierarchy, the effect of each fault mode is traced throughout the system, up to evaluate the reduction of performances and reliability/safety levels at the vehicle level. This analysis leads to a severity classification

Table 2.6 FMEA table template

Id. n.	Item	Function	Failure modes and causes	Mission phase/ operational mode	Failure effects			Failure-detection method	Compensating provision	Severity classification	Remarks
					Local effects	Next higher effects	End effects				
⋮	⋮	⋮	⋮	⋮	⋮	⋮	⋮	⋮	⋮	⋮	⋮
⋮	⋮	⋮	⋮	⋮	⋮	⋮	⋮	⋮	⋮	⋮	⋮

("Severity class" column in Table 2.7), which is given by a numerical code, ranging from I to IV to indicate catastrophic, hazardous, major, and minor consequences, respectively.

The FMEA also provides indications about the FDI capabilities of the system ("Failure detection method" column in Table 2.6). For complex airborne equipment (like EMAs), this field typically indicates the Built-in Test covering the fault.

A relevant example of CA application for airborne EMAs is given in [10], in which the PSSA activities for a MALE UAS FCS are addressed. As reported in Table 2.7, the CA on a simplex rotary EMA is developed by providing the fault severity classification and a Qualitative Probability of Occurrence (QPO, ranging from A to E to indicate probable, remote, extremely remote, and extremely improbable occurrences).

The CA results are summarized in terms of criticality matrix in Fig. 2.18, which provides clear indications on the priorities of corrective actions to be applied (the further along the diagonal line from the origin the fault mode is recorded, the greater the criticality). It is worth noting that, for a simplex rotary EMA, several corrective actions should be necessary to obtain a safety compliant equipment.

In [10], the safety concerns related to the application of a simplex EMA are overcome by a self-monitoring fault-tolerant design. In particular, the EMA is equipped with phase-isolating electronics, so that, in case of one fault of motor coils (EL03-05), current sensors (EL11-12), or MOSFET bridge (EC05), the control loops are reconfigured and the EMA operates without performance degradation with only two coils. In addition, the capability of the EMA to engage a fail-safe mode permits to cover the control electronics fault (EC06), the resolver and RVDT faults (EL08-10 and EL13-15), the power supply faults (EC01-04), wiring faults (EL16-17), and the magnet faults (EL06-07). All remaining faults can be then covered via maintenance (e.g.. all the jamming and structural failures) or by specific design solutions (e.g., standard rugged connectors).

Table 2.7 Criticality analysis of a simplex rotary EMA

Item	Fault mode	Fault ref.	Fault cause	QPO	Sev. class
Open circuit	Open circuit	EC01	Wire crack	D	I
	Short circuit	EC02	Wire chafing	D	I
			Insulation degradation		
	Thermal runway	EC03	Component dielectric breakdown	D	I
	Controller capacitor open/short circuit	EC04	Component dielectric breakdown	D	I
Power bridge	e.g. MOSFET open/short circuit	EC05	Various	C	I
Control electronics	e.g. CPU failure	EC06	Various	C	I
Motor connectors	Disconnection	EL01	Disconnection	E	I
	Contact degradation	EL02	Intermittent connection	D	I
Motor coil	Opened coil	EL03	Wire crack	C	III
	Shorted coil	EL04	Wire chafing	C	III
			Insulation degradation		
	Partially-shorted coil	EL05	Wire chafing	C	III
			Insulation degradation		
Motor magnet	Magnet separation	EL06	Bond degradation	E	I
	Magnet demagnetization	EL07	Thermal runway	C	III
Resolver	Opened coil	EL08	Wire crack	D	I
	Shorted coil	EL09	Wire chafing	D	I
			Insulation degradation		
	Disconnection	EL10	Disconnection	E	I
Current sensors	Opened coil	EL11	Wire crack	D	II
	Shorted coil	EL12	Wire chafing	D	II
			Insulation degradation		
RVDT	Opened coil	EL13	Wire crack	D	I
	Shorted coil	EL14	Wire chafing	D	I
			Insulation degradation		
	Disconnection	EL15	Disconnection	E	I
Wiring	Opened coil	EL16	Wire crack	D	I
	Shorted coil	EL17	Wire chafing	D	I
			Insulation degradation		
Four-bar linkage	Backlash	ME01	Wear	C	III
	Jamming	ME02	Wear	E	I
	Structural failure	ME03	Warping or load misalignment	E	I
Bearings	Backlash	ME04	Wear	C	III
	Jamming	ME05	Recirculating elements block	1	I
			Wear		
	Collapse	ME06	Warping or load misalignment	E	I

(continued)

Table 2.7 (continued)

Item	Fault mode	Fault ref.	Fault cause	QPO	Sev. class
Output lever	Structural failure	ME07	Crack	E	I
Gearbox	Backlash	ME08	Wear	C	III
	Jamming	ME09	Wear	E	I
	Collapse	ME10	Crack	E	I
Seals	Leakage	ME11	Wear	D	III
		ME12	Crack		
Lubricant	Loss of lubrication	ME13	Contamination	1	III
		ME14	Chemical breakdown		

Reproduced from [10]—originally published open access and licensed under CC-BY 3.0. https://journals.sagepub.com/doi/pdf/10.1177/1687814016644576

Figures 2.17 and 2.19 show the FTAs related to the total loss of the surface control with reference to the simplex EMA and to the self-monitoring fault-tolerant solution. The failure rate data used in the FTA are obtained from [1, 13, 18, 23] by assuming 55 °C operating temperature and Airborne Uninhabited Cargo (AUC) environment.

2.3.5 Built-in Tests

One of the most important functions required for safety-critical airborne systems is the capability to assess their operation state by performing automatic test procedures, called Built-in Tests (BIT), which are implemented in the control/monitoring software of the system. Depending on the fault criticality as well as on the tolerable latencies assigned to the fault, different types of BITs are foreseen:

- *Continuous Built-in Tests (CBITs)* are performed throughout the mission. They typically aim at detecting major fault modes characterized by constant failure rate (i.e., random regimes) and abrupt occurrence (e.g., electrical and electronic faults), so that the tolerable fault latencies are expected to be small;
- *Initialization Built-in Tests (IBITs)* or *Power-up Built-in Tests (PBITs)* are performed before starting the mission, immediately after that the system is powered. They aim at detecting all the major fault modes of the system, provided that the resulting duration of the test procedure is not excessive. Typically, the IBIT includes all CBIT checks and operates additional (detailed) monitoring algorithms;
- *Shut-down Built-in Tests (SBITs)* are performed at the end of the mission before the system is powered off. They aim at detecting all the major fault modes of the system, provided that the resulting duration of the test procedure is not excessive. Typically, the SBIT includes all CBIT checks and operates some additional monitoring algorithms (the total number of checks is less than the IBIT one);
- *Maintenance Built-in Tests (MBITs)* are performed on-ground, during the maintenance operations. They aim at detecting all the fault modes of the system and, thanks to less stringent requirements about the procedure duration, they can be suit-

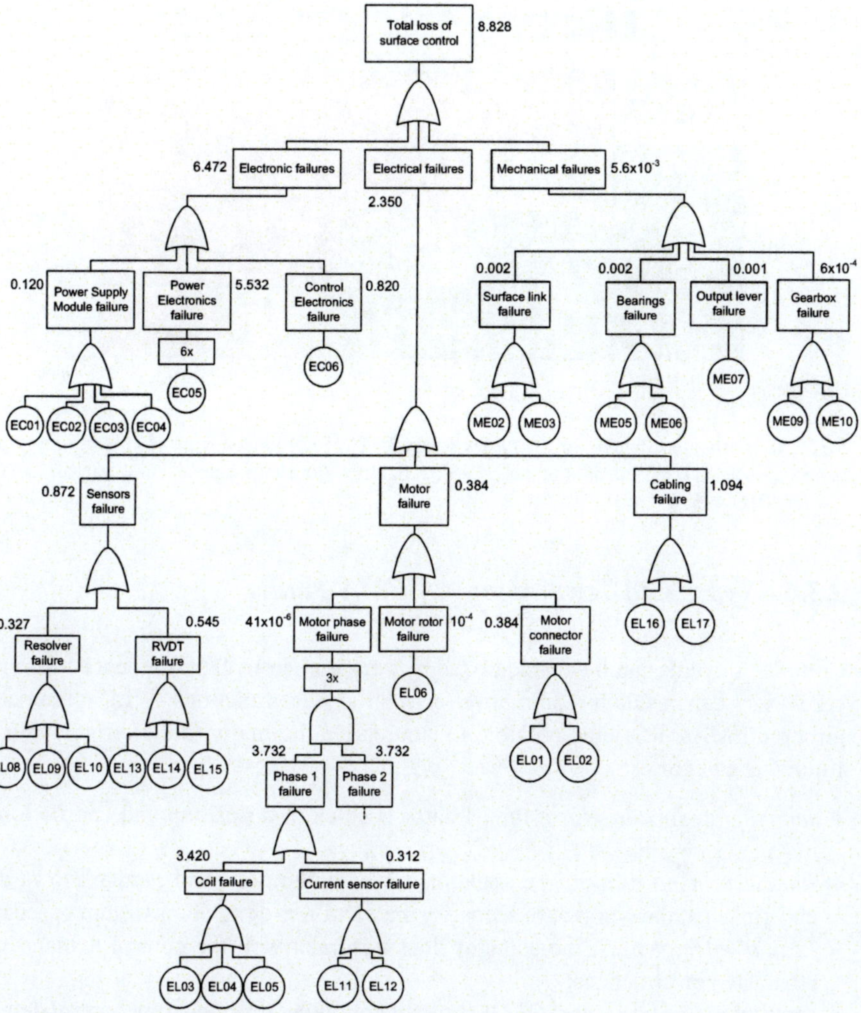

Fig. 2.17 FTA of a simplex flight control EMA (failure rates $\times 10^{-6}$ pFH). Reproduced from [10]—originally published open access and licensed under CC-BY 3.0. https://journals.sagepub.com/doi/pdf/10.1177/1687814016644576

able for the detection of fault modes characterized by failure rates that increase with the operative time (i.e., wear-out regimes) and by a slow evolution of the faulty behavior (e.g., mechanical faults). Typically, the MBIT includes all IBIT checks and operates additional (very detailed) monitoring algorithms.

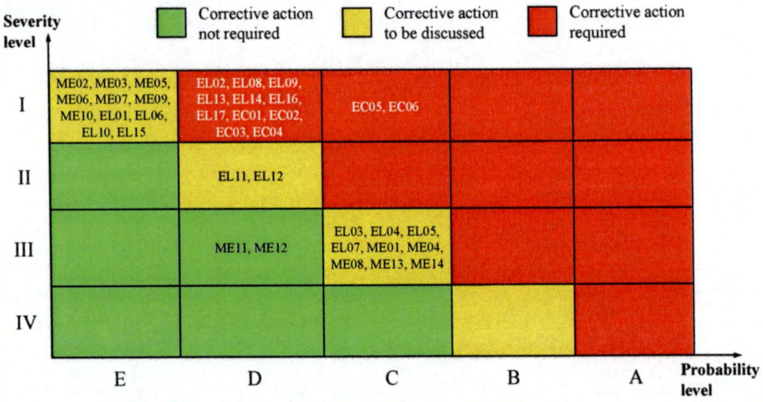

Fig. 2.18 Criticality matrix for a simplex rotary EMA. Reproduced from [10]—originally published open access and licensed under CC-BY 3.0. https://journals.sagepub.com/doi/pdf/10.1177/1687814016644576

2.3.6 Types and Terminology of EMA Faults

Complex systems can be characterised by very numerous fault modes, but in many cases they can result in common effects on system functions. With reference to airborne EMAs, it is thus possible to summarize the major faults' effects into the following categories:

- *Jamming* or *lock-in-place*: the actuator is stuck in a position and can no longer move;
- *Runaway* or *hard-over*: the actuator moves without control demand toward its endstroke (in the worst case scenario, the motion is done at maximum speed);
- *Loss of effectiveness*: the actuator does not track well the control demand with adequate performances;
- *Oscillatory Failure Case (OFC)*: the actuator output deviates from control demand by exhibiting abnormal oscillations;
- *Disconnection, free-play or free-floating*: the actuator connection with the load is not obtained as designed (diverging, small-amplitude and large-amplitude deviations imply disconnection, free-play or free-floating faults, respectively);
- *Stall*: the actuator dynamics is characterized by cyclic to intermittent saturation phenomena.

The jamming, runaway, loss of effectiveness, and OFC faults can be modeled following the rationale highlighted in Sect. 3.1.1 as (additive) *actuator faults*. The disconnection and stall faults can be modeled instead as *parametric faults*. For a list of electrical faults in EMAs, the reader can refer to Sect. 4.4.1.

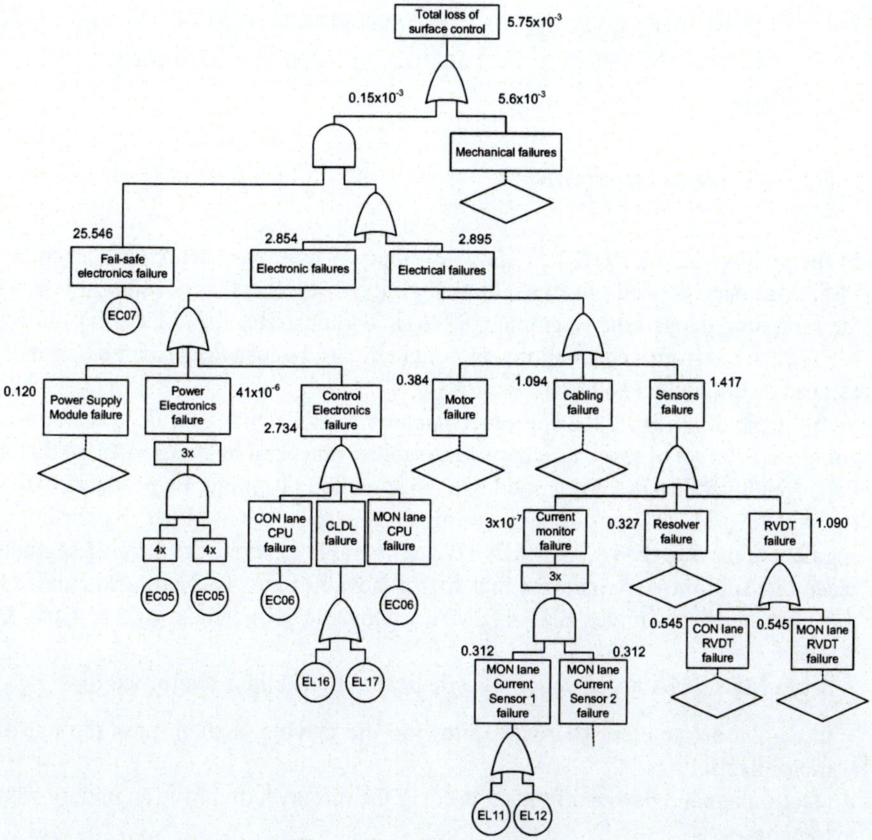

Fig. 2.19 FTA of a self-monitoring flight control EMA (failure rates $\times 10^{-6}$). Reproduced from [10]—originally published open access and licensed under CC-BY 3.0. https://journals.sagepub.com/doi/pdf/10.1177/1687814016644576

Fig. 2.20 Level of detail of system Built-in Tests (area is proportional to the number of tests)

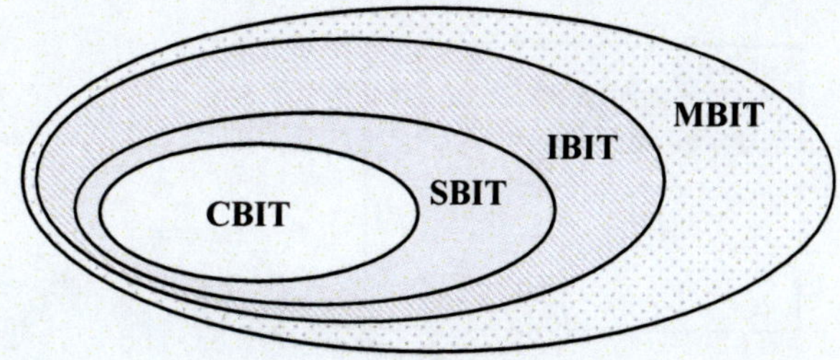

2.4 Preliminary System Safety Assessment of an Electro-Mechanical Actuation System for Morphing Flaps

2.4.1 System Description

In the context of the R&D activities developed within the CleanSky2 program, a PSSA has been carried out to evaluate the electro-mechanical actuation system of the morphing flaps of the Regional CS2 flight demonstrator [9, 22], i.e., a prototype MEA obtained from a conventional regional aircraft, by installing a number of more-electric technological solutions.

As depicted in Fig. 2.21, the outer flaps of the Regional CS2 demonstrator are composed of a set of seven articulated movables controlled by five EMAs, so that the flap is deployed in a morphing mode. In the so-called Morphing Flap System (MFS), two redundant EMAs in torque-summing arrangement are used to drive the central segment of the flap (EMA 1A, EMA 1B and Movable 1), while the three tip segments (each one split into two parts, so that six tip movables are obtained) are actuated by three independent simplex EMAs (EMA j, Movable jX where $j = 2, 3, 4$ and X = A, B).

Each MFS EMA is controlled by a dedicated ECU and is composed of

- three-phase frameless BLACM, allowing the driving shaft to pass through the motor itself;
- electromagnetic power-off brake, holding the motor shaft during stationary working phases;
- harmonic drive mechanical transmission [15] from motor to movable shaft;
- phase current sensors;
- motor resolver;

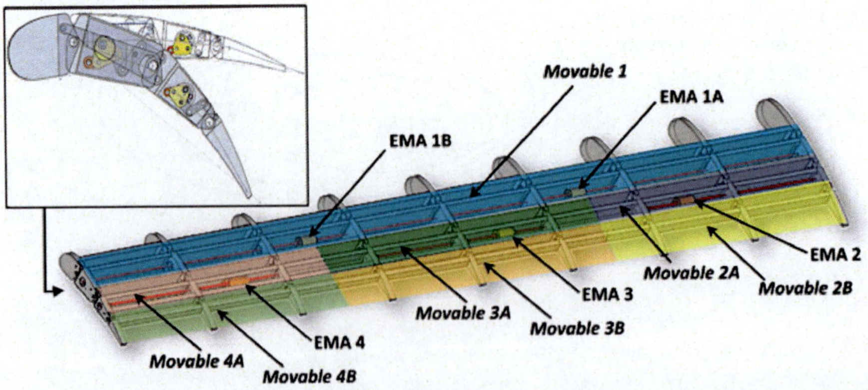

Fig. 2.21 Movables, mechanism and actuation system of the morphing flap (left wing of CS2 demonstrator)

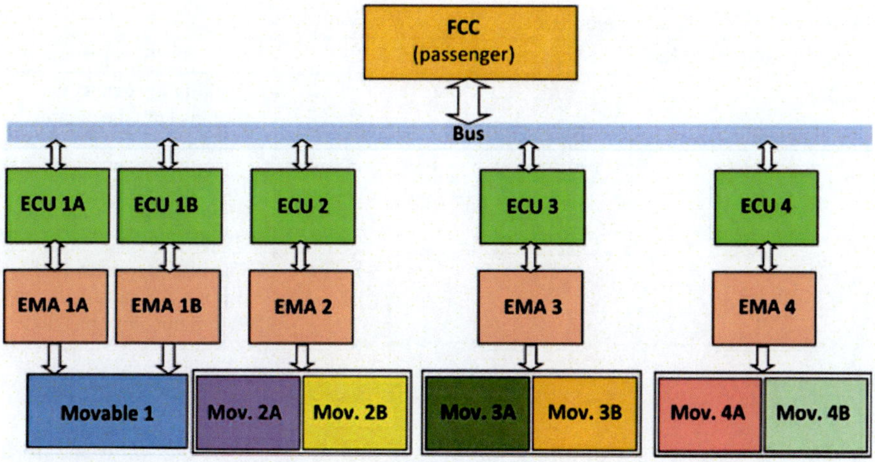

Fig. 2.22 MFS control architecture

- output position sensor.

The ECUs are connected to a passenger FCC (i.e., installed on the CS2 demonstrator as additional equipment, Fig. 2.22) and each one is essentially composed of:

- Power Supply Unit (PSU), to provide the electrical supply to the components (motor, brake, sensors);
- CPU, to implement the EMA closed-loop control laws;
- Power Electronics Unit (PEU), to implement the motor drive via SVPWM technique.

2.4.2 Operation Modes

The MFS operates during the flight mission by adapting the flap camber in order to

- enhance the high-lift performances and allow steeper initial climb and final descent trajectories, for noise-abatement purposes;
- enhance the wing aerodynamic efficiency in all flight conditions through the implementation of active load control functionalities.

The first task is implemented in the so-called *Mode 1*, in which all movables are simultaneously deployed to obtain large camber variations, while the second task can be accomplished in two modes: in the so-called *Mode 2*, only the tip segment movables are deployed all in the same direction to obtain a tab-like flap deflection; in the so-called *Mode 3*, only the outboard and inboard tip segments are deployed in opposite directions to obtain the twist of the flap trailing edge.

The MFS operative positions are (being a morphing surface, the deflection angles are considered as "equivalent" ones with respect to the neutral trailing edge shape):

Table 2.8 MFS functional requirements

Aircraft-level FHA Ref.	System-level FHA Ref.	System functional requirement
TBD	MFS1.1	Capability to connect the movables with the aircraft structure
TBD	MFS1.2	Capability to hold the movables in the operative position
TBD	MFS1.3	Capability to actuate the movables for transition between operative positions

- *Mode 1*: deployment of all movables to obtain up to 30° deflection (used during take-off, climb, descent, and landing phases);
- *Mode 2*: deployment of all tip segments in the same direction obtain up to ±10° deflection (used during all flight phases);
- *Mode 3*: differential deployment of inboard and outboard tip segments obtain up to ±5° trailing edge twist (used during all flight phases).

The stationary phases of operation are accomplished by engaging the brakes and by powering off the motors with related electronics. To obtain the transitions between the stationary positions, the EMA brakes are disengaged, the motor and the related power electronics are powered on, and the actuation power is regulated via position closed-loop control.

2.4.3 *Definition and Allocation of the Functional Requirements*

Table 2.8 defines the functional requirements of the left-wing MFS (MFS1), while Table 2.9 reports their allocation to MFS1 subsystems and components.

2.4.4 *Functional Hazard Analysis*

2.4.4.1 Functional Hazard Analysis Table

The FHA table for the MFS1 has been developed with reference to the following failure conditions:

1. total loss of function (i.e., no functionality);
2. partial loss (i.e., functionality is given with significant performance degradation),

Table 2.9 Functional requirements allocation to subsystems and components ($j = 2, 3, 4$)

MFS1 subsystem	Component	Functional requirement allocation		
		MFS1.1	MFS1.2	MFS1.3
Movable 1	Aerodynamic surface	x	x	x
	Bearings	x	x	x
	Mechanical joints	x	x	x
EMA 1A	Brake		x	x
	BLAC			x
	Current sensors			x
	Resolver			x
	Position sensor		x	x
	Harmonic drive gear		x	x
	Bearings		x	x
ECU 1A	PSU		x	x
	CPU		x	x
	PEU			x
EMA 1B	Brake		x	x
	BLAC			x
	Current sensors			x
	Resolver			x
	Position sensor		x	x
	Harmonic drive gear		x	x
	Bearings		x	x
ECU B	PSU		x	x
	CPU		x	x
	PEU			x
Movables *j*	Aerodynamic surface A	x	x	x
	Aerodynamic surface B	x	x	x
	Bearings	x	x	x
	Mechanical joints	x	x	x
EMA *j*	Brake		x	x
	BLAC			x
	Current sensors			x
	Resolver			x
	Position sensor		x	x
	Harmonic drive gear		x	x
	Bearings		x	x
ECU *j*	PSU		x	x
	CPU		x	x
	PEU			x

and by addressing the following mission phases:

(A) start-up/pre-flight;
(B) take-off/climb;
(C) cruise;
(D) descent/landing;
(E) shut-down/post-flight;
(F) maintenance.

As an example, the failure code "MFS1.2.1.C" stands for: Morphing Flap System 1, Functional requirement 2, Failure condition 1, Cruise. An excerpt of the FHA is reported in Tables 2.10 and 2.11.

2.4.4.2 Most Critical Failure Conditions

By analyzing the results of FHA table (Table 2.11), the Most Critical Conditions (MCC) related to the Left-Wing MFS are

MCC1 Loss of capability to connect one movable with the aircraft structure during take-off/climb, cruise or descent/landing phases FHA Ref. MFS1.1.2B-D. *Classification*: Hazardous/Catastrophic;

MCC2 Loss of capability to hold one movable in the operative position during take-off/climb, cruise or descent/landing phases FHA Ref. MFS1.2.3B-D. *Classification*: Hazardous/Catastrophic;

MCC3 Loss of capability to actuate one of the tip segment movables during take-off/climb, cruise or descent/landing phases FHA Ref. MFS1.3.5B-D. *Classification*: Hazardous/Catastrophic.

2.4.5 Fault-Tree Analysis

2.4.5.1 FTA of the Most Critical Failure Conditions

With reference to the MCC's selected from FHA tables, specific FTAs are developed and, by following a top-down workflow, the reliability requirements for MFS components are derived. It is worth noting that the indication of a reliability requirement in **bold** characters in an FTA means that the requirement is derived from that FTA. On the other hand, if a requirement is indicated in *italic* characters, it means that it has been derived from other FTAs. The FTAs related to the MFS MCCs are reported from Figs. 2.23, 2.24, and 2.25.

Table 2.10 Failure rate requirements definition for subsystems and components ($j = 2, 3, 4$)

MFS1 subsystem	Component	Failure rate requirement (pFH)	Criticality for simplex component (HIGH, Intermediate, Low)	Possible solutions to HIGH criticalities
Movable 1	Aerodynamic surface	n.d.	n.d.	
	Bearings	$<5 \cdot 10^{-10}$	Intermediate	
	Mechanical joints	$<5 \cdot 10^{-10}$	Intermediate	
EMA 1A	Brake	$<8 \cdot 10^{-11}$	**HIGH**	**Open point**
	BLAC	$<6 \cdot 10^{-6}$	Intermediate	
	Current sensors	$<1.95 \cdot 10^{-6}$	*Low*	
	Resolver	$<1.95 \cdot 10^{-6}$	**HIGH**	Dual redundancy
	Position sensor	$<1.95 \cdot 10^{-6}$	**HIGH**	Dual redundancy
	Harmonic drive gear	$<8 \cdot 10^{-11}$	**HIGH**	**Open point**
	Bearings	$<1.95 \cdot 10^{-11}$	**HIGH**	**Open point**
ECU 1A	PSU	$<1.95 \cdot 10^{-6}$	Intermediate	Dual redundancy
	PSUM	$<1.95 \cdot 10^{-6}$	Intermediate	
	PSUB	$<8 \cdot 10^{-11}$	**HIGH**	
	CPU	$<1.95 \cdot 10^{-6}$	Intermediate	
	PEU	$<1.95 \cdot 10^{-6}$	Intermediate	
EMA 1B	Brake	$<8 \cdot 10^{-11}$	**HIGH**	**Open point**
	BLAC	$<6 \cdot 10^{-6}$	Intermediate	
	Current sensors	$<1.95 \cdot 10^{-6}$	*Low*	
	Resolver	$<1.95 \cdot 10^{-6}$	**HIGH**	Dual redundancy
	Position sensor	$<1.95 \cdot 10^{-6}$	**HIGH**	Dual redundancy
	Harmonic drive gear	$<8 \cdot 10^{-11}$	**HIGH**	**Open point**
	Bearings	$<1.95 \cdot 10^{-11}$	**HIGH**	**Open point**
ECU 1B	PSU	$<1.95 \cdot 10^{-6}$	Intermediate	Dual redundancy
	PSUM	$<1.95 \cdot 10^{-6}$	Intermediate	
	PSUB	$<8 \cdot 10^{-11}$	**HIGH**	
	CPU	$<1.95 \cdot 10^{-6}$	Intermediate	
	PEU	$<1.95 \cdot 10^{-6}$	Intermediate	
Movables j	Aerodynamic surface A	n.d.	n.d.	
	Aerodynamic surface B	n.d.	n.d.	
	Bearings	$<5 \cdot 10^{-10}$	Intermediate	
	Mechanical joints	$<5 \cdot 10^{-10}$	Intermediate	

(continued)

Table 2.10 (continued)

MFS1 subsystem	Component	Failure rate requirement (pFH)	Criticality for simplex component (HIGH, Intermediate, Low)	Possible solutions to HIGH criticalities
EMA *j*	Brake	$<6 \cdot 10^{-6}$	*Low*	
	BLAC	$<2 \cdot 10^{-6}$	Intermediate	
	Current sensors	$<6.5 \cdot 10^{-7}$	Intermediate	
	Resolver	$<6.5 \cdot 10^{-7}$	**HIGH**	Dual redundancy
	Position sensor	$<6.5 \cdot 10^{-7}$	**HIGH**	Dual redundancy
	Harmonic drive gear	$<3 \cdot 10^{-6}$	*Low*	
	Bearings	$<3 \cdot 10^{-6}$	*Low*	
ECU *j*	PSU	$<6.5 \cdot 10^{-7}$	Intermediate	
	PSUM	$<6.5 \cdot 10^{-7}$	Intermediate	
	PSUB	$<10^{-9}$	**HIGH**	Dual redundancy
	CPU	$<6.5 \cdot 10^{-7}$	**HIGH**	Dual redundancy
	PEU	$<6.5 \cdot 10^{-7}$	**HIGH**	Dual redundancy

2.4.5.2 Failure Rate Requirements for Subsystems and Components

A summary of the FTA results in terms of failure rate requirements for the subsystems and components is reported in Table 2.10, together with an evaluation of the criticality of the requirements with respect to typical failure rates of airborne components (Sect. 2.1.1).

It can be noted that there are several criticalities and, for some cases, a redesign of the system architecture is required (see "Open point" Table 2.10). Actually, for the MFS EMAs related to tip segments (EMAj with $j = 2, 3, 4$) the use of a dual redundancy in electric and electronic components can lead to a flightworthy solution. On the other hand, for the MFS EMAs related to the central segments (EMA1A and EMA1B), the basic reliability concern is related to mechanical parts, due to the single load path of the actuation.

2.5 Summary

In the first part of the chapter, the basic reliability/safety concerns related to the airworthiness certification of airborne systems (like EMAs) are addressed. Starting from the discrepancy between the typical failure rates of subsystems/components and the reliability/safety levels required for certification, different redundancy paradigms are presented (load-level, system-level, and subsystem-level redundancy) and the most

Table 2.11 Excerpt of the MFS FHA table

Failure code	Function	Phase	Failure condition	Failure effects	Classification	Remarks/mitigating actions
MFS1.1.2.A	Capability to connect the movables with the aircraft structure	Start-up/Pre-flight	Mechanical disconnection of one Left-Wing MFS movable (partial loss)	The motion of the disconnected movable is out of control	Major	Failure can be detected either visually by ground operators or via IBIT
				The actuation of one movable is not possible		
MFS1.1.2.B	Same as MFS1.1.2.A	Take-off/ Climb	Same as MFS1.1.2.A	The motion of the disconnected movable is out of control	Hazardous/Catastrophic	Failure can be detected via CBIT
				Free-floating of the movable causes aeroelastic phenomena		
				The movable is not held in the correct operative position		
				The actuation of one movable is not possible		
MFS1.1.2.C	Same as MFS1.1.2.A	Cruise	Same as MFS1.1.2.A	Same as MFS1.1.2.B	Same as MFS1.1.2.B	Same as MFS1.1.2.B
MFS1.1.2.D	Same as MFS1.1.2.A	Descent/Landing	Same as MFS1.1.2.A	Same as MFS1.1.2.B	Same as MFS1.1.2.B	Same as MFS1.1.2.B
MFS1.1.2.E	Same as MFS1.1.2.A	Shut-Down/ Post-Flight	Same as MFS1.1.2.A	Same as MFS1.1.2.A	Minor	Failure can be detected either visually by ground operators or via SBIT
MFS1.1.2.F	Same as MFS1.1.2.A	Maintenance	Same as MFS1.1.2.A	Same as MFS1.1.2.A	Minor	Failure can be detected either visually by operators or via MBIT

(continued)

Table 2.11 (continued)

Failure code	Function	Phase	Failure condition	Failure effects	Classification	Remarks/mitigating actions
MFS1.2.3.A	Capability to hold the movables in the operative position	Start-up/Pre-flight	Loss of Left-Wing MFS capability to hold one movable in the operative position (partial loss). Possible failures: Undetected runaway of one movable, Free floating of one movable	The movable drifts away from the correct deflection	Major	Failure can be detected either visually by ground operators or via IBIT. Considering that visual detection is possible only with relevant external loads, a dedicated IBIT check is foreseen (e.g., the motor is activated with engaged brake, and the hold-on capability is checked). Dormant failure can result in catastrophic effect during flight
				Limit cycle motion occurs when the movable is actuated The movable is out of control		
MFS1.2.3.B	Same as MFS1.2.3.A	Take-off/ Climb	Same as MFS1.2.3.A	The movable drifts away from the correct deflection Limit cycle motion occurs when the movable is actuated The movable is out of control Free floating of the movable causes aeroelastic phenomena	Hazardous/Catastrophic	Failure can be detected via CBIT
MFS1.2.3.C	Same as MFS1.2.3.A	Cruise	Same as MFS1.2.3.A	Same as MFS1.2.3.B	Same as MFS1.2.3.B	Same as MFS1.2.3.B
MFS1.2.3.D	Same as MFS1.2.3.A	Descent/ Landing	Same as MFS1.2.3.A	Same as MFS1.2.3.B	Same as MFS1.2.3.B	Same as MFS1.2.3.B
MFS1.2.3.E	Same as MFS1.2.3.A	Shut-Down/ Post-Flight	Same as MFS1.2.3.A	Same as MFS1.2.3.A	Minor	Failure can be detected either visually by ground operators or via SBIT
	Same as MFS1.2.3.A	Maintenance	Same as MFS1.2.3.A	Same as MFS1.2.3.A	Minor	Failure can be detected either visually by operators or via MBIT

(continued)

Table 2.11 (continued)

Failure code	Function	Phase	Failure condition	Failure effects	Classification	Remarks/mitigating actions
MFS1.3.5.A	Capability to actuate the movables for transition between operative positions	Start-up/Pre-flight	Loss of Left-Wing MFS capability to actuate one of the tip segment movables (partial loss of Mode 2 or total loss of Mode 3) Possible failures (worst cases): Jamming of the movable at ±10°, Loss of function of the movable at ±10°, Uncontrolled motion	One of the tip segment movables is blocked in neutral position	Major	Failure can be detected either visually by ground operators or via IBIT.
				One of the tip segment movable is out of control (e.g., not completely deflected, in oscillating motion)		
MFS1.3.5.B	Same as MFS1.3.5.A	Take-off/Climb	Same as MFS1.3.5.A	One of the tip segment movables is blocked in position and does not respond to motion commands One of the tip segment movables is out of control (e.g., not completely deflected, in oscillating motion)	Hazardous/ Catastrophic	Failure can be detected via CBIT
MFS1.3.5.C	Same as MFS1.3.5.A	Cruise	Same as MFS1.3.5.A	Same as MFS1.3.5.B	Same as MFS1.3.5.B	Same as MFS1.3.5.B
MFS1.3.5.D	Same as MFS1.3.5.A	Descent/Landing	Same as MFS1.3.5.A	Same as MFS1.3.5.B	Same as MFS1.3.5.B	Same as MFS1.3.5.B
MFS1.3.5.E	Same as MFS1.3.5.A	Shut-Down/ Post-Flight	Same as MFS1.3.5.A	Same as MFS1.3.5.A	Minor	Failure can be detected visually by ground operators or via SBIT
MFS1.3.5.F	Same as MFS1.3.5.A	Maintenance	Same as MFS1.3.5.A	Same as MFS1.3.5.A	Minor	Failure can be detected visually by ground operators or via MBIT

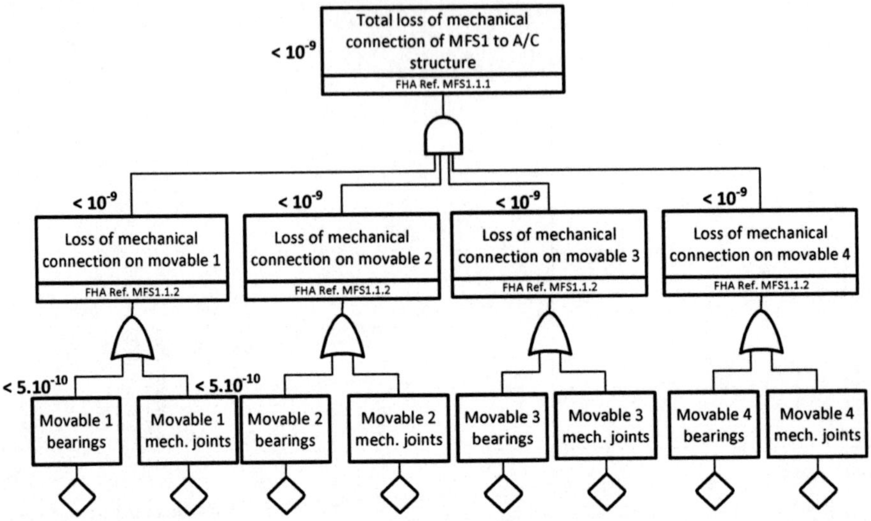

Fig. 2.23 FTA related to the loss of mechanical connection capability

Fig. 2.24 FTA related to the loss of the "hold in position" capability

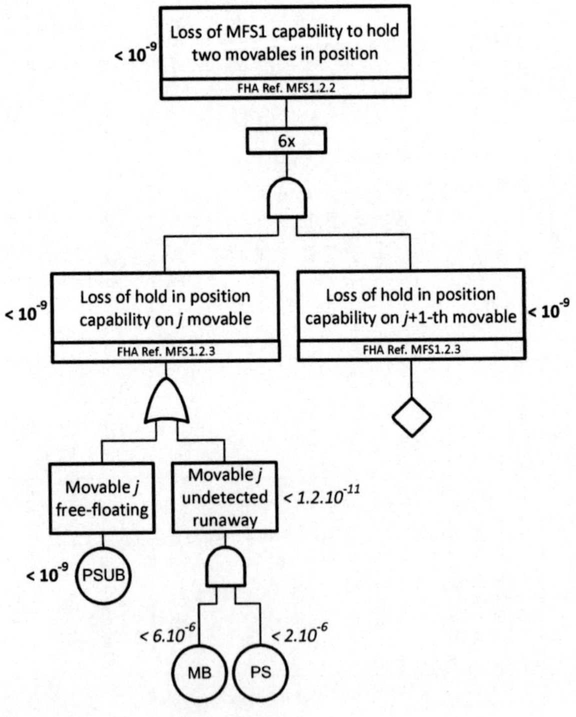

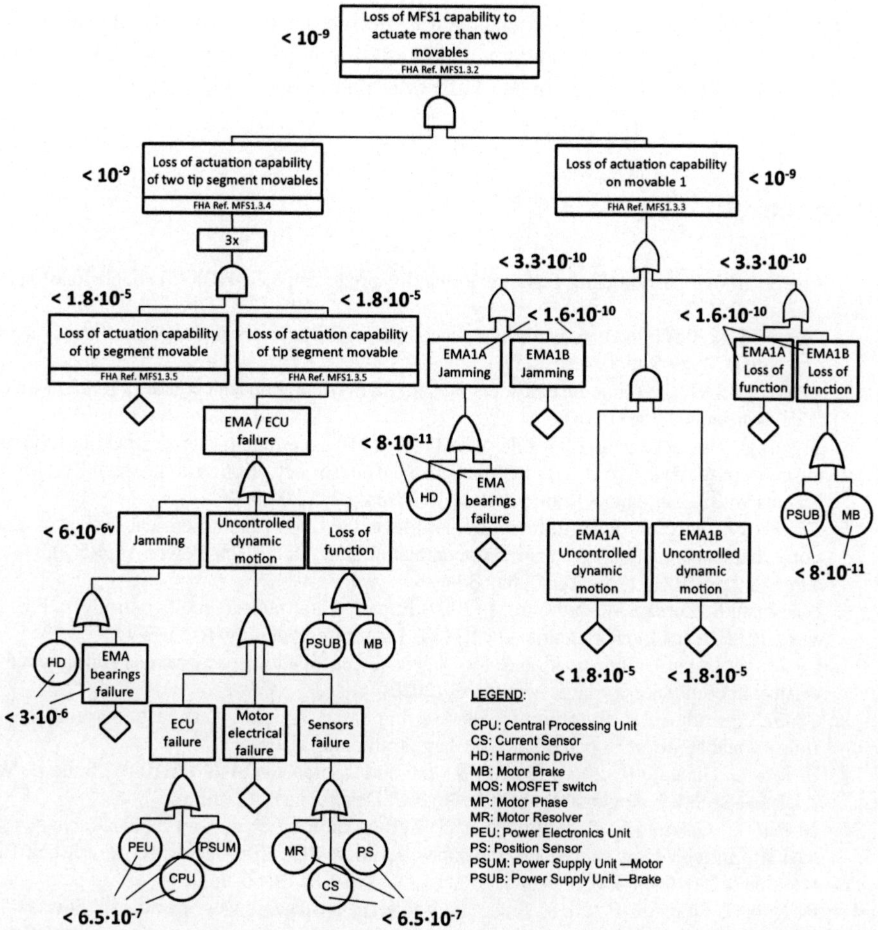

Fig. 2.25 FTA related to the loss of actuation capability

relevant design solutions for fault-tolerant EMAs are described (redundant power electronics, motors with redundant phrases, jamming-tolerant mechanical transmission). The standardized guidelines, methods, and analyses for the safety assessment of airborne system are successively discussed, by also introducing practical examples of FTA and FMECA for aircraft EMAs. A specific section is then dedicated to the definition and the categorization of system built-in tests (CBIT, IBIT, SBIT, and MBIT).

In the second part of the chapter, the PSSA carried out to evaluate the EMA system of the morphing flaps of the Regional Clean Sky 2 flight demonstrator is presented and discussed. Starting from the description of the system architecture and operation modes, the functional requirements are defined and their allocation to subsystems and components is illustrated. An excerpt of the FHA table referred to the most critical failure condition is reported and the related FTAs are developed. Finally, by following a top-down workflow, the FTAs are used to derive the reliability requirements for subsystems and components, and an evaluation of the criticality of the requirements with respect to typical failure rates of airborne components is proposed.

References

1. American National Standards Institute: Reliability prediction MIL-HBK-217F subsidiary specification (2013)
2. Annaz FY (2005) Fundamental design concepts in multi-lane smart electromechanical actuators. Smart Mater Struct 14(6):1227
3. ARP4754, SAE: Certification considerations for highly-integrated or complex aircraft systems. SAE, Warrendale, PA (1996)
4. Arriola D, Thielecke F (2017) Model-based design and experimental verification of a monitoring concept for an active-active electromechanical aileron actuation system. Mech Syst Signal Process 94:322–345. https://doi.org/10.1016/j.ymssp.2017.02.039
5. Bennett JW, Mecrow BC, Atkinson DJ, Atkinson GJ (2011) Safety-critical design of electromechanical actuation systems in commercial aircraft. IET Electr Power Appl 5(1):37–47. https://doi.org/10.1049/iet-epa.2009.0304
6. Bolognani S, Zordan M, Zigliotto M (2000) Experimental fault-tolerant control of a PMSM drive. IEEE Trans Indus Electron 47(5):1134–1141. https://doi.org/10.1109/41.873223
7. CS-25, European Union Aviation Safety Agency: Certification specifications and acceptable means of compliance for large aeroplanes (2020)
8. CS-29, Amendment 7, European Union Aviation Safety Agency: Certification specifications and acceptable means of compliance for large rotorcraft (2019)
9. Di Rito G, Galatolo R (2018) FHA PSSA method assessment, AG2 O1.10-01, Issue D, WP 2.1.1, Clean Sky 2, Regional Innovative Aircraft Development Platform
10. Di Rito G, Galatolo R, Schettini F (2016) Self-monitoring electro-mechanical actuator for medium altitude long endurance unmanned aerial vehicle flight controls. Adv Mech Eng **8**(5):1687814016644,576. https://doi.org/10.1177/1687814016644576
11. Di Rito G, Schettini F (2018) Smart fault-tolerant air-data sensor for aircraft flow angles measurement. In: 31st congress of the international council of the aeronautical sciences (ICAS), pp 1–7
12. Guo H, Xu J, Kuang X (2015) A novel fault tolerant permanent magnet synchronous motor with improved optimal torque control for aerospace application. Chinese J Aeronaut 28(2):535–544. https://doi.org/10.1016/j.cja.2015.01.008
13. Handbook, Military: Electronic reliability design handbook. In: MIL-HDBK-338, DoD (1988)
14. Hardier G, Seren C, Ezerzere P, Puyou G (2013) Aerodynamic model inversion for virtual sensing of longitudinal flight parameters. In: 2013 conference on control and fault-tolerant systems (SysTol), pp 140–145. https://doi.org/10.1109/SysTol.2013.6693835
15. Harmonic Drive mechanical transmissions. https://www.harmonicdrive.net/
16. Jiang R, Murthy D (2011) A study of Weibull shape parameter: properties and significance. Reliab Eng Syst Saf 96(12):1619–1626. https://doi.org/10.1016/j.ress.2011.09.003
17. Manohar GA, Vasu V, Srikanth K (2018) Development of a high redundancy actuator with direct driven linear electromechanical actuators for fault-tolerance. Procedia Comput Sci **133**:932–939 (2018). https://doi.org/10.1016/j.procs.2018.07.089. International Conference on Robotics and Smart Manufacturing (RoSMa2018)
18. MIL-HDBK-217F-Notice 2. Reliability prediction of electronic equipment. https://snebulos.mit.edu/projects/reference/MIL-STD/MIL-HDBK-217F-Notice2.pdf
19. O'Connor P, Kleyner A (2012) Practical reliability engineering. Wiley. https://doi.org/10.1002/9781119961260

20. Oosterom M, Babuska R (2000) Virtual sensor for fault detection and isolation in flight control systems—fuzzy modeling approach. In: Proceedings of the 39th IEEE conference on decision and control (Cat. No.00CH37187), vol 3, pp 2645–2650 (2000). https://doi.org/10.1109/CDC.2000.914204
21. Oosterom M, Babuska R, Verbruggen HB (2002) Soft computing applications in aircraft sensor management and flight control law reconfiguration. IEEE Trans Syst Man Cybern, Part C (Appl Rev) **32**(2):125–139. https://doi.org/10.1109/TSMCC.2002.801357
22. Rea F, Amoroso F, Pecora R, Noviello MC, Arena M (2018) Structural design of a multifunctional morphing fowler flap for a twin-prop regional aircraft. https://doi.org/10.1115/SMASIS2018-7937. V001T04A003. Paper ID: SMASIS2018-7937
23. Reliability information Analysis Center (RiAC): NPRD-2011 (2011)
24. Roma (Italy): Ministero della Difesa – Direzione Generale degli Armamenti Aeronautici: AER(EP)-P. 6, Istruzioni per la compilazione dei capitolati tecnici per aeromobili militari. (2012). http://nso.nato.int/nso/nsdd/stanagdetails.html?idCover=7520&LA=EN
25. RTCA DO-178C, Radio Technical Committee Association: Software considerations in airborne systems and equipment certification (2012)
26. RTCA DO-254: Design assurance guidance for airborne electronic hardware (2012)
27. ARP4761, SAE: Guidelines and methods for conducting the safety assessment process on civil airborne systems and equipment. SAE International (1996)
28. Schettini F, Di Rito G, Denti E, Galatolo R (2017) Wind identification via kalman filter for aircraft flow angles calibration. In: 2017 IEEE international workshop on metrology for aeroSpace (MetroAeroSpace), pp 97–102 (2017). https://doi.org/10.1109/MetroAeroSpace.2017.7999545
29. Schettini F, Di Rito G, Galatolo R (2018) Smart air-data probe for fault-tolerant flow measurements. In: 2018 5th IEEE international workshop on metrology for aeroSpace (MetroAeroSpace), pp 602–607. https://doi.org/10.1109/MetroAeroSpace.2018.8453555
30. Schettini F, Di Rito G, Denti E (2019) Aircraft flow angles calibration via observed-based wind estimation. Aircraft Eng Aerosp Technol. https://doi.org/10.1108/AEAT-06-2017-0145
31. Yu ZY, Niu T, Dong HL (2018) A jam-tolerant electromechanical system. In: ACTUATOR 2018; 16th international conference on new actuators, pp 1–4

Chapter 3
Fault Diagnosis and Condition Monitoring Approaches

Outline of the chapter. This chapter presents the concepts of fault diagnosis and condition monitoring that will be used in the remainder of the book. Special attention is posed to giving a clear and concise definition of the different concepts and terms of interest. Furthermore, the various approaches that have been developed to achieve monitoring and diagnosis goals are discussed.

Section 3.1 introduces the basic definitions in the context of fault diagnosis and condition monitoring, giving a classification of the different fault types based on their location or on their time behavior. Section 3.2 reviews common diagnostic methodologies, while Sect. 3.3 focuses especially on application of these methodologies to electro-mechanical actuators. Finally, Sect. 3.4 summarizes the chapter.

3.1 Basic Concepts and Terminology

Depending on the consequences related to the loss of operating functions, engineering systems can be safety critical. Illustrative cases are chemical and industrial processes [66, 71], power networks [13], electric machines and components [9], wind energy conversion systems [78], power plants [27], manufacturing lines (machines, tools, robots, transportation systems), and several other examples [45]. For safety-critical systems, it is of paramount importance to immediately uncover potential abnormalities of the underlying system or process. Components that possess nonstandard conditions may result sooner or later in undesirable effects. These effects, in turn, can exhibit different behaviors, as specified in the following definitions.

3.1.1 *Fault, Failure, Malfunction, Disturbance, Model Uncertainty*

A *fault* is generally defined as an unpermitted deviation of at least one characteristic property (feature) or parameter of the system from the acceptable/usual/standard

© Springer Nature Switzerland AG 2021

M. Mazzoleni et al., *Electro-Mechanical Actuators for the More Electric Aircraft*, Advances in Industrial Control, https://doi.org/10.1007/978-3-030-61799-8_3

condition, due to an unexpected event (e.g., physical component failure or supply breakdown). In particular, the following hypotheses apply:

- the unacceptable violation is relative to certain specification limits (tolerance zones) for normal operation;
- a fault can be present even if the system is not running;
- when in operation, a fault may not affect the correct functioning of a system, i.e., the system can still be able to perform the function for which it has been designed.

Modeling of faults, disturbances, and model uncertainties. Faults can be mathematically modeled as additional external signals or as parameter deviations [84]. In the first case, the faults are called *additive faults*. They are represented by an unknown input that enters the model equations as an addend. In the second case, the faults are called *multiplicative faults*. Here, the magnitude of the fault depends on the magnitude of some system control input. Sensor faults and several types of actuator faults are usually considered as additive faults, while parametric faults (abnormal variation of some model parameters) are considered as multiplicative faults. An alternative approach is called *physical fault modeling*. It involves multiple models, representing a collection of individual models where each model corresponds to a specific fault situation (e.g., changing the model parameters, adding/removing a specific signal).

A *disturbance* $d(t)$ is an nonmeasurable uncontrolled input acting on the system. Disturbances are usually represented by additional input signals to be superimposed, from the system inputs up to the system output, like additive faults. A relevant example of disturbances are the loads due to wind turbulence acting on an aircraft or external loads acting on a plant.

A *noise* input $w(t)$ is another form of nonmeasurable uncontrolled input acting on the system. Differently with respect to a disturbance, for noise inputs, we do not know how it affects the system output (e.g., we do not know the transfer function from $w(t)$ to $y(t)$). Thus, in some fault diagnosis designs, it is possible to completely reject the effect of disturbance on the quantities used for fault diagnosis, but the noises effects on them can only be attenuated [84]. Some authors, e.g., [11, 22], make no difference between disturbance and noise inputs, while defining the fault diagnosis problem with Linear Time-Invariant (LTI) dynamic models, while other authors explicitly consider both disturbances and noises in their formulation [34, 84].

Model uncertainties are, given a fixed model structure, possible changes to the model parameters. In this regard, they affect the model in a similar way as multiplicative faults.

However, an important distinction between disturbances, model uncertainties, and faults can be seen in the fact that disturbances and model uncertainties are always present, while faults may be present or not. A general schematic of the signals involved in an open-loop plant is shown in Fig. 3.1. Here, the main entities are

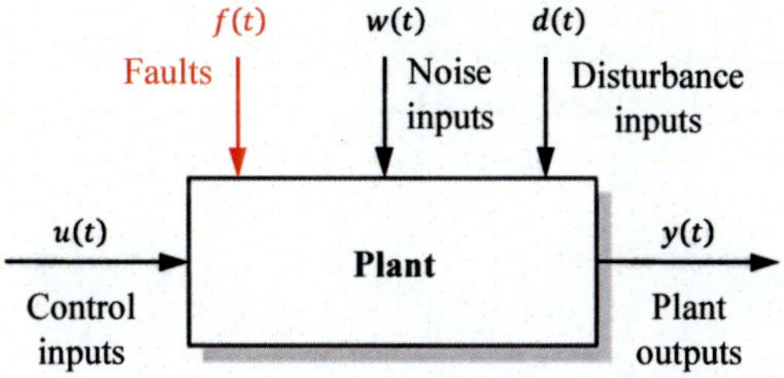

Fig. 3.1 Plant model with faults, where u is the control input, y is the plant output, w represents an unknown noise source or parametric uncertainties, and d is a physical disturbance

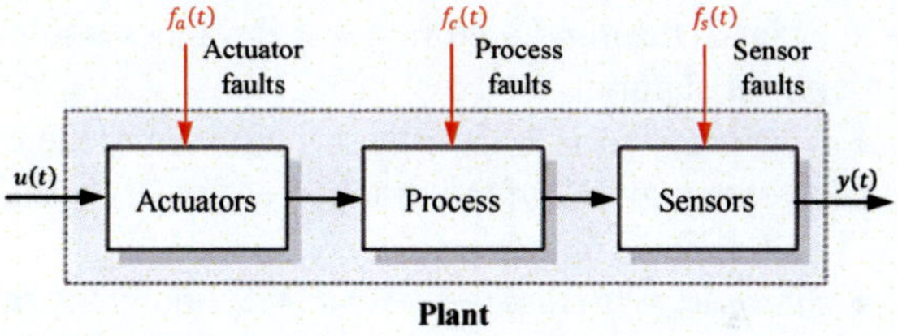

Fig. 3.2 Distinction between actuator faults f_a, process/component faults f_c and sensor faults f_s, for a system with input u and output y

- the control input $u(t)$;
- the output measurement $y(t)$;
- disturbance input $d(t)$;
- the noise input $w(t)$;
- the fault input $f(t)$.

Notice that we can have multiples of each one of these variables. Usually, $u(t)$ and $y(t)$ are the only measurable variables, and they have to be used to build the diagnosis and monitoring algorithm. Typical noise inputs are sensor noise signals as well as process input noise; however, noise inputs can represent also uncertainties on the model parameters.

Classification of faults. Faults are often classified by considering the "location" where they affect the system, see Fig. 3.2:

- *process (plant, component, parameter) faults*: imply a modification in the dynamic behavior of the system and they can cause a variation of the input/output relationship;
- *sensor faults*: imply an incorrect measurement of the sensor readings;
- *actuator faults*: imply an undesired modification of the control actions, which can be interrupted or modified.

Due to the location of different faults in the cause-effect chain of the plant, there are specific methods for detecting and counteracting them [4].

Another classification concerns time scales characterizing the transition from normal to faulty behavior, with respect to those of the system itself, so they may arise

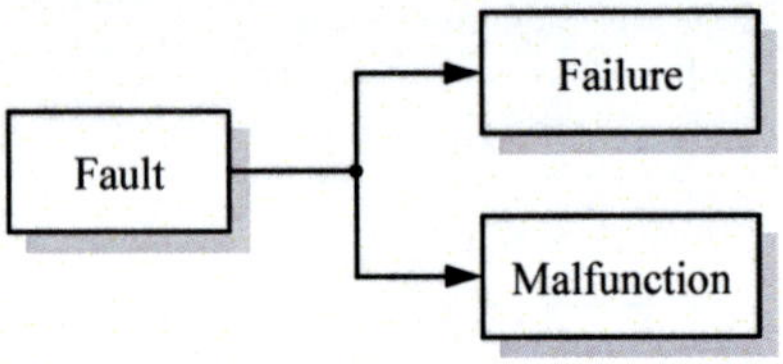

Fig. 3.3 Development of the events "failure" and "malfunctions" from a fault

- *abruptly* (step wise): these faults are modeled as stepwise functions and represent a bias in the monitored signal. Relevant examples is the breakdown of an electrical power supply;
- *incipiently* (drift wise): these faults are modeled using ramp functions and they represent a drift of the monitored signal. A relevant example is the wearing of a mechanical component;
- *intermittently* (impulse wise): these faults are modeled by pulse signals with different amplitudes. A relevant example is an intermittent electrical contact.

Faults versus Failures and Malfunctions. It is important to differentiate the meanings of faults, failures, and malfunctions. Both failures and malfunctions originate from one or more faults, and they usually arise after the beginning of operation or by increasingly stressing the system, see Fig. 3.3.

A *failure* is a permanent interruption of a system capability to perform a required function under specified operating conditions. In particular [44]:

- a failure is an event, and it can be represented by a Boolean logic quantity;
- the definition of failure applies also where a functional unit of the system is not able to perform its duty [43];
- failures can be distinguished according to
  - *Predictability*:
    - (i) random: i.e., failures that are unpredictable (statistically independent from operating time or other failures);
    - (ii) deterministic: i.e. some failure features can be predicted by observing certain conditions (there is a correlation between failure and observations);
    - (iii) systematic or causal: i.e., failures that are dependent on known conditions (there is a causality relation between failure and observations).

A *malfunction* is an intermittent irregularity in the fulfillment of a system desired function. The only difference with a failure is that, in the malfunction case, the interruption of the system function is only temporary. The degradation of the system performance (that can be stable over time), can be also thought of as a malfunction.

3.1.2 Fault Diagnosis, Condition Monitoring, and Fault Prognosis

Dependability refers to a general aggregate of system qualities such as reliability, availability, and safety, so that a dependable system is a *fail-safe* system (i.e., having the capability to respond to a failure by reverting to a safe passivation/shutdown with no, or minimal, harm to other equipment, environment, or persons) with high availability and reliability.

Dependability is among one of the most critical issues in the design of today's automatic control systems. A traditional way to improve system dependability is to enhance the quality and robustness of system parts like sensors, actuators, controllers, or computers. Even so, a fault-free system operation cannot be guaranteed. Fault diagnosis and condition monitoring are thus introduced to detect unwanted situations in the monitored system.

Fault Diagnosis. The concept of *fault diagnosis* refers to the general usage of specific techniques to assess the status of a system with respect to its possible faults. In this book, we consider fault diagnosis as entailing the following essential tasks:

1. *Fault detection*: to discover anomalous behaviors occurring in the functional units of the plant. It consists in the *detection* of occurrence of faults and the determination of the *time* at which the fault occurs.
2. *Fault isolation*: (following the fault detection) to locate a fault within the plant, i.e., the *localization* of detected faults.
3. *Fault estimation*: (following the fault isolation) to *reconstruct* the time-varying behavior (shape) of the fault signals.
4. *Fault analysis* or *identification*: (following the fault isolation) to characterize the *type, size (severity),* and *nature (cause)* of detected faults.

Sometimes, the terms *fault estimation* and *fault identification* are used as synonyms, where one includes also the aims of the other. Unfortunately, the terminology in the field is not consistent, especially in the coverage of the term *fault diagnosis*, since it often depends on research context, application, and publication period. The definitions of the concepts, aims, and procedures for fault diagnosis and monitoring slightly varies across the research communities, such as

- process industry and statistical process monitoring community [66, 68, 71];
- early automatic control and fault diagnosis community [11, 46, 73, 77];
- current automatic control and fault diagnosis community [4, 22, 23, 31, 44, 84].

This book adheres to the nomenclature used by the current trends in the control community, see also Fig. 3.4.

The aforementioned tasks can be regarded in terms of increasing complexity. Fault Detection (FD) systems are the simplest fault diagnosis schemes. FD systems trigger an alarm signal to indicate the *presence of any fault* in the system, without telling which fault (or multiple faults) occurred. The decision on the occurrence or absence

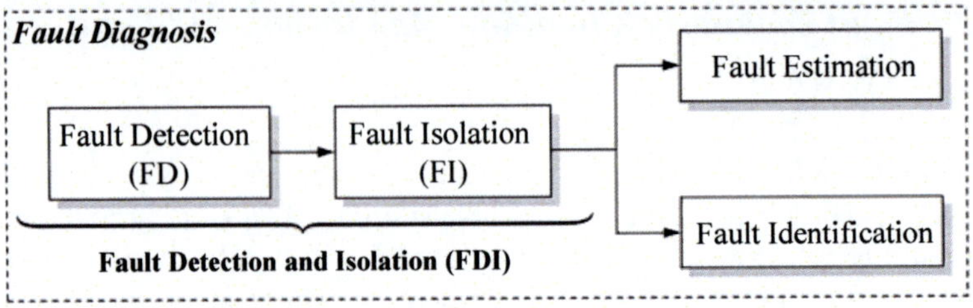

Fig. 3.4 Basic steps involved in fault diagnosis

of faults must be done in the presence of arbitrary control inputs, disturbance inputs, and noise inputs acting simultaneously on the system.

Fault Detection and Isolation (FDI) or Fault Detection, Isolation and Analysis (FDIA) systems deliver exact localization of the occurred faults. Roughly speaking, the decision on the presence or absence of each considered fault must be taken. This means to *isolate one fault from another*. Ideally, such decisions must be achieved regardless of the faults occur one at a time or several faults occur simultaneously. One can distinguish between *strong fault isolation* if an arbitrary number of multiple occurring faults can be isolated, and *weak fault isolation* where a limited number of simultaneous faults has to be assumed. Fault estimation, following fault isolation, tries to *reconstruct the fault signals* from the available measurements. Fault estimation can serve as a basis for control law reconfiguration or virtual sensor development, see Sect. 3.1.3. Fault identification, usually subsequent to fault isolation, is used to fully characterize the *type, size,* and *nature* of the faults.

Assessing the importance of different tasks is problem dependent. Fault detection is the basic level of utility for any practical system and isolation is almost equally important. Fault estimation and identification, on the other hand, may not be essential if no reconfiguration action is required. For these reasons, fault diagnosis is very often considered as fault detection and isolation in the literature.

Condition monitoring. With the term *condition monitoring*, we refer to the *continued oversight* of the progression of the degradation of a system or a component. The aim is to design one or more indicators of the state of health of the monitored object. These indicators can be directly defined by a specific property of the system, as in cases where there is a causal relation between indicator and system health, or it can be devised in such a way that the indicator is only correlated with the system health state.

Fault diagnosis versus condition monitoring. The main difference between fault diagnosis and condition monitoring lies in how the output of the methods is treated. In the fault diagnosis case, we are usually interested in a logical Boolean output or *dichotomous answer*, i.e., the presence or absence of a fault (and where it is located). Fault diagnosis techniques are able to provide deep insights on what faults are present, what components are faulty, and what is the entity of the faults. In condition monitoring, instead, we look for a *continuous evaluation* of the system condition. Ideally, condition monitoring aims at generating an indicator of health state that monotonically evolves as system/component degradation progresses. Usually, this progression manifests with the lifetime increase. Condition monitoring algorithms can be applied system wise or component wise. In the first case, a compound assessment of the health state is given. In the second case, various components have to be monitored, usually with a different and specific indicator for each component. This translates, normally, into using an almost different set of sensors for each component. Generally speaking, if multiple faults are present, fault diagnosis methods provide a more structured approach for accomplishing its tasks, in a way that a single framework is required to manage multiple faults. Condition monitoring, on the other hand, needs to adapt to the different components monitored in a specific manner.

The distinction between fault diagnosis and condition monitoring is not so sharp, and sometimes the two terms are used equivalently. The two methodologies share common approaches to produce specific responses. Condition monitoring can be accomplished by the electronics embedded in a sensor or an actuator by measuring the deviation of a self-monitored variable. Fault diagnosis, instead, requires a more structured approach to perform not only the detection, but also the isolation and the estimation/identification stages. In the following, we will assume that the techniques defined for fault diagnosis are also valid for condition monitoring.

Fault prognosis. With *fault prognosis*, we denote the procedures that collect the continuous indications of the current health state of the system (provided by the condition monitoring function), to forecast its future evolution. The prognosis of a fault usually requires the development of a model for the evolution of the fault, in order to extrapolate the future fault trend. Fault prognosis is an iterative procedure, since the predictions have to be updated every time a new "health state point" is generated.

Fault diagnosis versus condition monitoring versus fault prognosis. In the specific view of the authors, the following differences appear when the terms fault diagnosis, condition monitoring, and fault prognosis are used. As previously specified, we distinguish mainly based on the type of the output given by the method, and based on which time instant that output is referred. Fault diagnosis outputs a *dichotomous* output at each *actual time*. The same applies for condition monitoring, with the exception of producing a *continuous* output. Fault prognosis is built on condition monitoring, by further providing values for the monitoring indicator (or indicators) that are referred to *future times*.

Figure 3.5 provides a graphical representation of these concepts. The graph plots a degradation index as a function of lifetime. Blue dots represent the value given by the condition monitoring indicator, and the red dotted horizontal line is the fault threshold. At time $t = 7$, there is the exceeding of the threshold, and the fault diagnosis (fault detection in this case) indicator goes from low to high state, indicating the detected presence of a fault. The prognosis module, a time $t = 3$, predicts that the threshold will be exceeded at time $t = 6$. With more available data points, the prognostic model updates its predictions for the evolution of the degradation index.

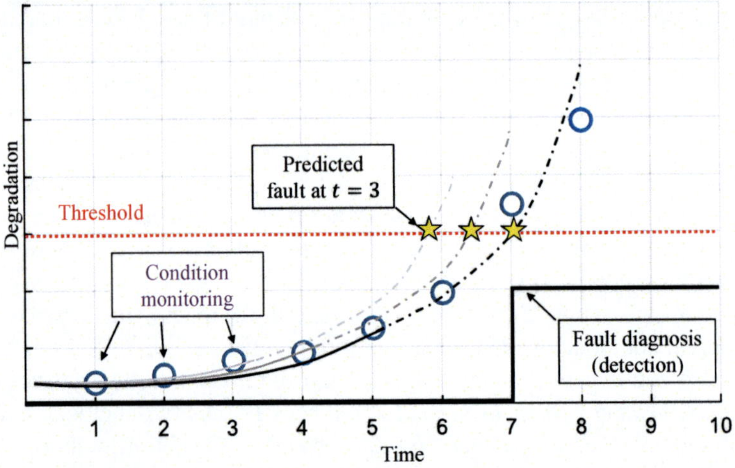

Fig. 3.5 Intuition about the different scopes of fault diagnosis, condition monitoring, and fault prognosis. (Blue dots) condition monitoring indicators; (black line) fault detection decision; (red dotted line) fault threshold; (gray lines) prognostic model and predictions

3.1.3 Fault-Tolerant Systems

Faults, and what follows from them, could cause product deterioration or damage to machines and persons. A *fault-tolerant* system is able to hinder the propagation of these detrimental effects [4]. Fault-Tolerant Control (FTC) undertakes measures such that the system function is satisfied even also after the appearance of faults. FTC has to prevent a component fault from causing a failure at the system level. In case of failure, the aim is also to reconfigure the controller and control system so that a fail-safe operation of the automatic control system is guaranteed. It is possible to distinguish between *active fault-tolerant control* and *passive fault-tolerant control*. The first methodology consists of a *supervision system* that, reacting to the diagnosed occurrence of a fault, performs fault accommodation or control reconfiguration, as shown in Fig. 3.6. *Fault accommodation* deals with the autonomous adaptation of the controller parameters to the faulty plant behavior. *Control reconfiguration* includes the selection of a new control configuration. *Reconfigurability* is a property of the plant that stands for the possibility of finding a new controller that satisfies the control aims for the faulty system. A fault-tolerant controller that performs fault accommodation does not modify the structure of the control laws, but adapts its parameters to the faulty behavior of plant, in order to meet acceptable performances. Passive fault-tolerant control happens when the control loop tolerates faults, with small effects on the plant, thanks to its robustness. For these reasons, a fault-tolerant closed-loop system *could hide* the presence of a fault to an external observer, because the system still satisfies its designated goals even when the fault is present. Finally, we talk about *fault compensation* as the process of actively intervening to modify the hardware/software architecture of the system after a fault,

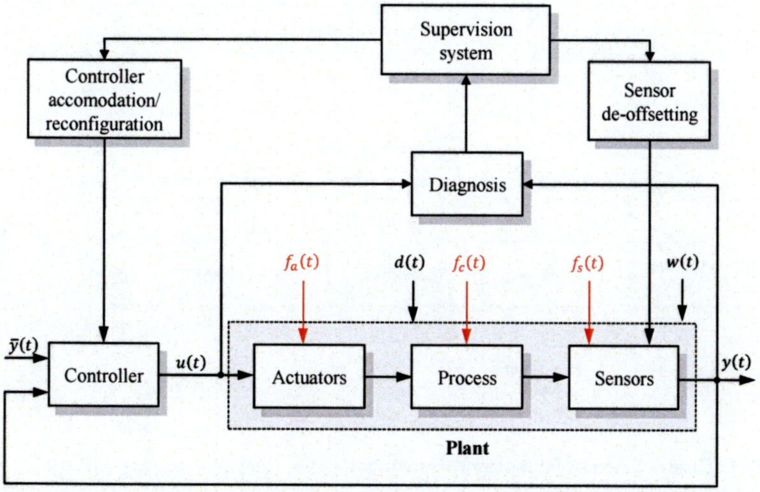

Fig. 3.6 Architecture of a fault accommodation and control reconfiguration scheme, with control action u, output y, reference $\bar{y}$, disturbance d, and faults f_a, f_c, f_s. For a non-recoverable fault, the supervision level has to make a decision about the system objectives (e.g., safe shutdown), since the current objectives can no longer be achieved

aiming to recover some level of system performance. Contrary to fault accommodation, where no component is turned off, here the control law is changed and also the faulty component is deactivated.

Fault-tolerant control is the immediate step following a successful fault diagnosis, with particular reference to the stages after fault detection. The locations of the faulty components and their damage levels described by the types, shapes, and sizes of the faults are vital for the system to take responsive fault-tolerant actions and to remove the detrimental effects that the faulty parts have on the system's normal operation. The supervision system can not only reconfigure the controller. Faults on sensors can be managed by appropriately offsetting the faulty signals by actuator/sensor signal compensation, and even replacing faulty components with redundant duplicates, in a way that the adverse effects from faults are accommodated or removed.

3.2 Common Diagnostic Methodologies

In this section, we provide a taxonomy for the different methodologies that can be employed to achieve fault diagnosis and condition monitoring of a system or a component, see Fig. 3.7. For a description of hardware and software redundancy on airborne EMAs, see Sects. 2.1.3 and 2.1.4.

Hardware redundancy. The basic idea underlying the concept of fault diagnosis is the concept of *redundancy*. This basically means one or more replications of

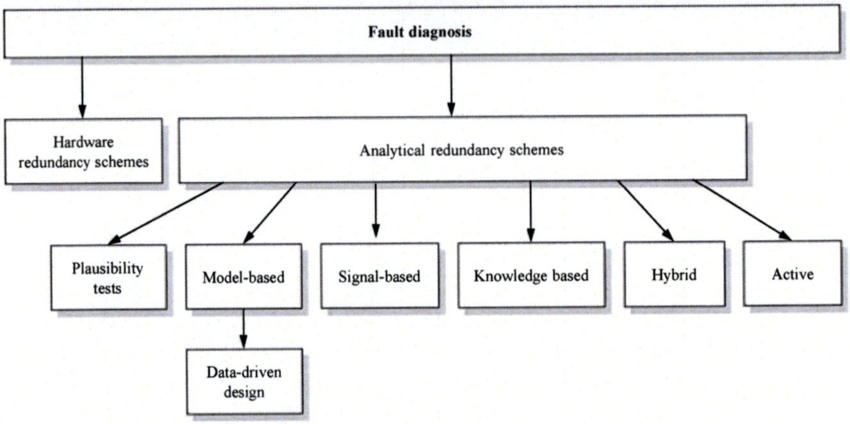

Fig. 3.7 Classification of fault diagnosis methods

the critical components or functions of a system. A traditional approach to fault diagnosis is based on the concept of *hardware* (or physical/parallel) *redundancy*. These methods use multiple lanes of sensors, actuators, computers, and software to measure and/or control a particular variable. In many safety-critical systems, such as Fly-By-Wire systems in aircraft, some parts of the control system may be triplicated. This is generally called Triple Modular Redundancy (TMR). Then, a voting scheme is applied to the hardware redundant system to decide if and when a fault has occurred and its likely location. The basic idea of hardware redundancy is to use identical components with the same input signal so that the duplicated output signals can be compared, leading to a diagnostic decision by a variety of methods such as limit checking and majority voting.

Hardware redundancy is reliable but suffers from major drawbacks. These problems consist of the extra equipment and maintenance cost and the additional space required to accommodate the equipment. For these reasons, hardware redundancy would not be applicable if it is applied to the whole system.

Analytical redundancy. Modern control theory paved the way for the so-called *analytical redundancy schemes*. As the name suggests, these methods still rely on the concept of redundancy. However, redundancy is no more achieved by using a physical duplicate of the system. Instead, the inputs u and outputs y, for the controlled system subjected to actuator fault f_a, process/component fault f_c, and sensor fault f_s, are employed to construct a fault diagnosis algorithm. These computations extract information that characterizes the system's health state. The extracted features are compared against a previous knowledge of the healthy system. Previous knowledge can be in the form of a physical or a black-box model, known patterns and behaviors of the signals inside the system, or even automatically learned thresholds and rules from historical data. Finally, a diagnostic decision is made, see Fig. 3.8.

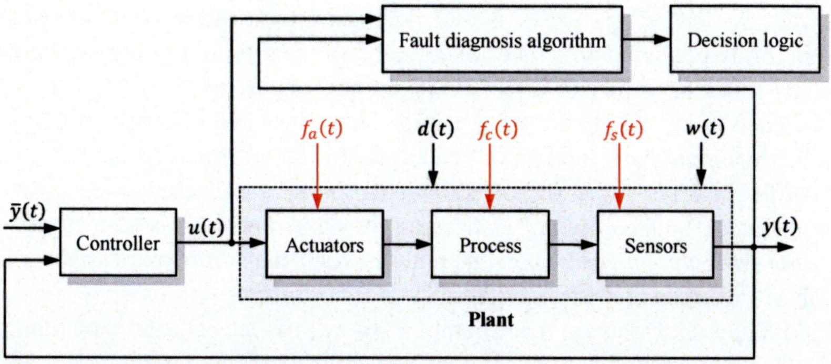

Fig. 3.8 Analytical redundancy-based fault diagnosis

Compared with hardware redundancy methods, analytical redundancy diagnostic methods are more cost-effective, but more challenging due to environmental noises, inevitable modeling errors, and the complexity of the system dynamics and control structure.

Taxonomy of analytical fault diagnosis methods. The concept of analytical redundancy can be implemented in many different ways. The aim is to check for consistency of real-time processed data and previous knowledge of how the considered computed features should behave in case the system was healthy (i.e., without faults). For these reasons, fault diagnosis methods based on the analytical redundancy framework can be classified into the following approaches [23, 31, 32]:

- *Plausibility test approach*: the plausibility test is based on the check of some simple physical laws under which a process component works. It is assumed that if there is a fault, this will lead to loss of plausibility. It consists of a set of rules that define the boundary values for a set of interrelated variables.
- *Model-based approach*: the core of the model-based approach is the development of an analytical process model that describes the process dynamics and its major features. A model-based fault diagnostic system consists of two parts: (i) *generation* of a so-called *residual signal*, which is generated by processing the system inputs and outputs; (ii) *residual evaluation* and decision-making. The model-based methods are powerful in dealing with fault diagnosis in dynamic processes.
- *Signal-processing-based approach*: these methods lie on the assumption that certain process signals carry information about the faults to be detected. The output of the system is collected and some indicators are computed. This indicators are also called *fault symptoms*. There is knowledge of the value of the symptoms in the healthy state. Fault diagnosis is achieved by comparing the current value of the symptoms with their values in the healthy state. Signal-processing-based methods are mainly used for steady-state processes.

- *Knowledge-based approach*: knowledge-based fault diagnosis is based on the availability of a great amount of historical process data. In this case, there is not *a priori* behavior or pattern to be compared with actual information, as in the case of model-based and signal-based approaches. The a priori information is hidden in the historical data and has to be extracted. This information can be leveraged in different ways. However, all knowledge-based approaches share a two-stage procedure: (i) a *training phase*, where previous knowledge is condensed; (ii) an *online evaluation phase*, where the real-time data are compared with the information extracted from the training phase.
- *Hybrid approach*: this approach combines more than one of the previous approaches, with the aim of taking the benefits of each considered methodology.
- *Active approach*: this method actively injects a specifically designed control input into the process, in order to allow easier detection of existent faults.

3.2.1 Model-Based Approach

The works of [10, 90] lied the first key concepts for model-based analytical redundancy. Model-based fault diagnosis can be defined as *the determination of faults of a system from the comparison of available system measurements with a priori information, provided by the system mathematical model, through the generation of residual quantities and their analysis* [11]. In model-based methods, the model of the industrial process is necessary. This can be obtained by using either physical principles or systems identification techniques. The model acts as a *digital twin*: the system behavior is reproduced by its mathematical counterpart. Using the same inputs, the measured outputs and the outputs predicted by the model are compared, Fig. 3.9. The difference between the measured process variables and their estimates is called a *residual*. The residual signal carries the most important message for a successful fault diagnosis.

The procedure of creating the residual signal is called *residual generation*. The process model and the comparison unit form the so-called *residual generator*. The residual generator can be thought of as a filter that, processing the inputs $u(t)$ and outputs $y(t)$ of the system, generates the residuals $r(t)$. The residuals are then evaluated (by computing some norm or by statistical hypothesis techniques), in order to be comparable with a defined threshold. Then, a diagnostic decision $\iota(t)$ is taken by comparing the evaluated residuals $\theta(t)$ with the threshold. The residual generator should be designed to be robust not only with respect to the possible system inputs, but also decoupled from the disturbances $d(t)$. In this way, the residuals will be sensible only to the presence of the faults.

Residual generation can be thought of as an extended plausibility test, where the consistency is based on the system model. On the basis of the type and application of the system model, it possible to group the model-based methods into four categories [31]:

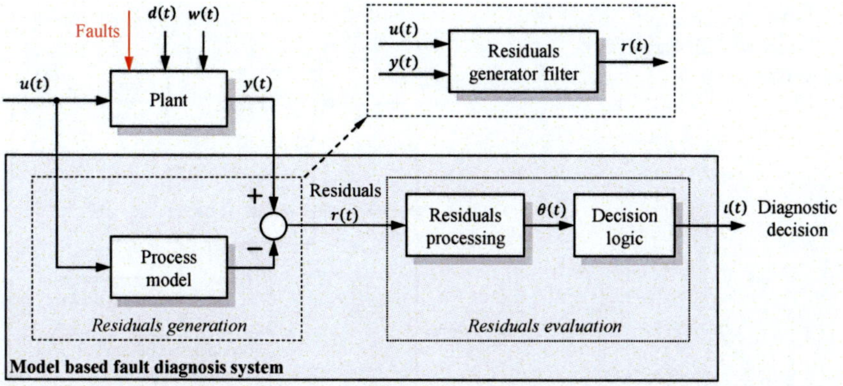

Fig. 3.9 Schematic description of the model-based fault diagnosis scheme. Adapted by permission from Springer Nature. Model-Based Fault Diagnosis Techniques Design Schemes, Algorithms and Tools by Steven X. Ding © (2013)

1. deterministic fault diagnosis methods;
2. stochastic fault diagnosis methods;
3. discrete events and hybrid systems methods;
4. networked and distributed systems methods.

An alternative taxonomy can be based on the level of complexity of the physical description characterizing the system models, see e.g., Sects. 4.3.1 and 4.3.3:

1. reduced-order models;
2. high-fidelity models.

3.2.1.1 Deterministic Fault Diagnosis Methods

The following techniques are used for fault diagnosis of systems characterized by a deterministic model:

- observers schemes;
- parity relations;
- stable coprime factorization.

Observer schemes. Observer plays a key role in model-based fault diagnosis for monitored systems/processes characterized by deterministic models, Fig. 3.10. In this approach, an observer is placed in the "process model" block of Fig. 3.9. Observers require that the model structure and parameters are known, and employ an output error correction scheme to track the system state variable. However, one of the aims of the residual generator is also to achieve a decoupling of the fault of interest from other faults, unknown disturbances, and model uncertainties.

The basic idea behind the development of the observer-based fault diagnosis technique is

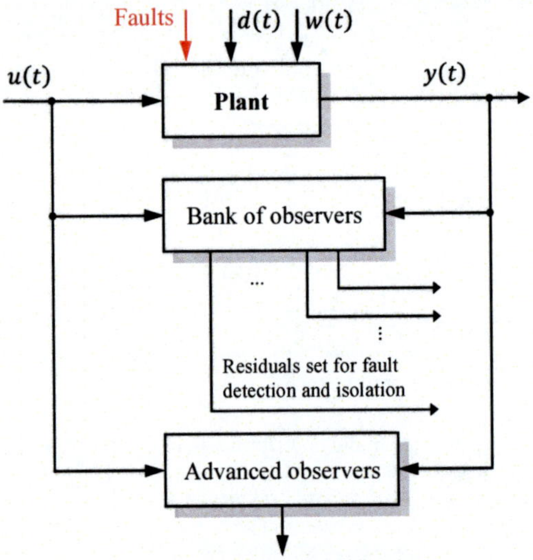

Fig. 3.10 Schematic of model-based fault diagnosis based on observers

- to replace the process model by an observer which will deliver reliable estimates of the process outputs;
- to provide the designer with the needed design freedom to achieve the desired decoupling using the well-established observer theory.

The design of the observer gain matrix can be tackled via an eigenstructure assignment [63] or via Linear Matrix Inequality (LMI) [50]. A bank of observer-based residuals is generally required in order to accomplish *fault isolation*, see Fig. 3.10. We can distinguish between

- *structured residual fault isolation*: a single residual is sensitive to the fault concerned, but robust against other faults, disturbances, and modeling errors [35];
- *generalized residual fault isolation*: each residual signal is sensitive to all but one fault and robust against modeling errors and disturbances.

The Unknown Input Observer (UIO) [12] is an observer scheme that can be used for fault isolation due to its ability to decouple the residual, corresponding to a specific fault, from input disturbance, modeling errors, and other faults.

It is important to highlight that there is a difference between the unknown input observer scheme (used to estimate the state of a dynamic system) and the unknown input residual generation scheme (used to generate residuals for fault diagnosis). In fact, the core of an observer-based residual generator is an output observer whose existence conditions are different (less strict) from those for a (state) unknown input observer. Furthermore, the unknown input decoupling scheme only focuses on the unknown inputs without explicitly considering the faults. As a result, the unknown input decoupling is generally achieved at the cost of the fault detectability [22].

Advanced observer techniques such as Proportional and Integral (PI) observers [93], Proportional Multiple-Integral (PMI) observers [49], adaptive observer [36], sliding mode observers [1], and descriptor observers [33] are usually utilized for fault estimation and fault identification. Here the idea is is to construct an augmented system by introducing the concerned fault as an additional state, and the extended state vector is then estimated. Therefore, the advanced observers are also called *simultaneous state and fault observers*.

Parity equations. The parity equations approach consists of directly comparing the outputs of the model with the respective measured outputs. In this approach, a model of the system is placed in the "process model" block of Fig. 3.9. This model is used to generate residuals, also called *parity vectors* in this approach. Both state-space and input/output models (transfer functions) can be employed. The difference with the observer-based method lies in the fact that there is not an output error correction scheme. One of the significant properties of parity-relation-based residual generators, also widely viewed as the main advantage over the observer-based approaches, is that the design can be carried out in a straightforward manner. In fact, it only deals with solutions of linear equations or linear optimization problems. However, the similarities between parity equations and observer-based scheme are deep and well investigated in the literature. In particular, the parity space methods lead to certain types of observer structures (such as a dead-beat observer) even though the design procedures differ. For this reason, the two approaches are usually treated together. A common strategy based on *"parity space design, observer-based implementation"* is vastly used: first, design the parity vector by linear algebra techniques; then, use the parity vector to build a diagnostic observer.

The advantages of this rationale to design a model-based diagnostic system lie in the facts that

- it is easier to design a parity vector rather than directly the observer;
- the observer configuration scheme is online (closed-loop configuration), while the parity space approach requires to store several previous inputs and outputs values (open-loop configuration) [22].

The parity relation approach can be applied to either time domain or frequency domain.

Stable coprime factorization. The stable factorization approach is a frequency-domain fault diagnosis method [24]. The main rationale is to generate a residual based on the stable coprime factorization of the transfer function matrix of the monitored system. The obtained residual can be made sensitive to the faults, but robust against disturbances, by selecting an optimal weighting factor. The stable fractional approach has connections with observers since the method includes the design of the observer gain, together with the state-feedback gain.

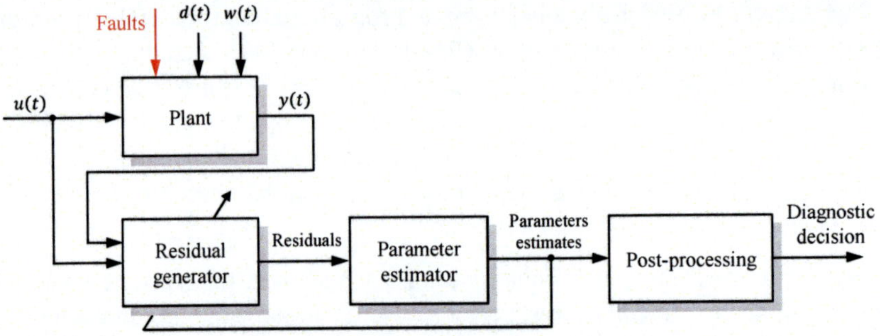

Fig. 3.11 Schematic description of the parameter identification scheme. Adapted by permission from Springer Nature. Model-Based Fault Diagnosis Techniques Design Schemes, Algorithms and Tools by Steven X. Ding © (2013)

3.2.1.2 Stochastic Fault Diagnosis Methods

The following techniques are used for fault diagnosis of systems characterized by a stochastic model:

- Kalman and particle filters;
- parameters estimation.

In the following, we will briefly describe those techniques.

Kalman and particle filters. Fault diagnosis methods based on Kalman Filtering (KF) are the ideal counterpart of the observers' schemes for stochastic systems. Here, the residuals (or innovations) generated by Kalman filters are looked for whiteness, mean, and covariance changes. Techniques for evaluating these residuals are based on generalized likelihoods, χ^2 testing, cumulative sum algorithms, and multiple hypothesis test. Extension of the basic Kalman filter to nonlinear systems such as the Extended Kalman Filter (EKF) and the Unscented Kalman Filter (UKF) were employed for effective fault diagnosis [29]. Adaptive Kalman filters (that can tune the process noise covariance matrix) and augmented state Kalman filters (able to estimate the fault as an additional state) are also used. Furthermore, the Particle Filter (PF), a Sequential Monte Carlo (SMC) method that approximates the state distribution via a set of particles, is also effectively used for fault diagnosis [61].

Parameters estimation. In the parameters estimation framework, the faults are assumed to be reflected in system parameters. Contrary to the observer and parity relation schemes that require a known model structure and parameters, here only the model structure is needed to be known. A fault decision is based on an online parameter estimation, Fig. 3.11. The estimation is performed by system identification techniques [77]. The estimated parameters are then compared with the reference parameters obtained initially under healthy conditions. These methods are very straightforward if the model parameters have an explicit mapping with the physical coefficients.

Efforts have been made to compare observer schemes with the parameters estimation one. Both schemes have advantages and disadvantages in different aspects, and there are arguments for and against each scheme. For instance, observers and parity equation methods are more suited for additive faults, while parameters estimation schemes are more prone to detect multiplicative faults.

An interesting union of the two approaches led to the *adaptive observers*. The major difference between the adaptive observer-based and parameter identification schemes lies in the residual generation.

3.2.1.3 Data-Driven Design of Model-Based Fault Diagnosis Methods

The so-called Stable Kernel Representation (SKR) of Linear Time-Invariant (LTI) systems offers an explicit method to define a unified representation for all types of LTI model-based residual generator schemes [22]. The SKR is a stable linear system in which expression can be derived from the known state-space representation of the process.

Recently, studies on Subspace Identification Methods (SIM)-based FDI system design drawn remarkable research attention, see e.g., [26]. The basic idea behind these schemes is a *direct construction of an FDI system utilizing the collected process data without explicitly identifying a system model*. We call these schemes *direct data-driven design of analytical fault diagnosis systems*. This rationale differs from standard SIM: in that case, the task is to identify the system model and the associated system matrices. Based on this, a residual generator is built using the SKR obtained from identified system matrices. Instead, the idea of direct data-driven FDI methods is to *directly identify from data* the SKR, upon which the residual generator is devised. Advantages of such schemes are

- simplified design procedures;
- ability to deal with dynamic processes.

In this way, the system identification becomes a part of the FDI system design and implementation procedure. Applications of these methods can be found in [20, 21, 23, 25].

3.2.1.4 Fault Diagnosis for Discrete Events and Hybrid Systems

In this section, we review the approaches to fault diagnosis for discrete events and hybrid systems.

Discrete event systems. In discrete event systems, the signals do not evolve continuously but change from one value to another in a discrete manner. Two main approaches exist:

- automata-based methods;
- Petri net-based methods.

Automata-based methods, in order to reduce the complexity of the task, gave birth to different solutions: (i) decentralized methods [64]; (ii) symbolic methods [74]; (iii) a combination of decentralized and symbolic methods [38].

Petri nets, thanks to their intrinsically distributed nature, possess an asset to reduce the computational complexity of solving fault diagnosis problems [8].

Hybrid systems. Systems where continuous and discrete dynamics coexist are called *hybrid systems*. The two dynamics are mutually dependent and interact with each other. The most common model, used to represent hybrid systems, is the *hybrid automata* approach. These models can be employed to design fault diagnosis algorithms to detect and isolate faults [96].

Another approach is to use *Bond graphs*. From them, it is possible to obtain parity relations from the causalities on the graph. A recent result of Bond-graph-based fault diagnosis and their applications to hybrid systems can be found in [52].

3.2.1.5 Fault Diagnosis for Networked and Distributed Systems

In this section, we review the approaches to fault diagnosis for networked and distributed control systems.

Networked systems. Real-time control and monitoring via communication channels is called *networked control and monitoring*. In network-based fault diagnosis, in addition to modeling errors, process disturbance, and measurement noises, the residual signal has to be robust also to transmission delays, data dropouts, and incomplete measurements caused by the limited capacity of communication channels. An example is given in [40], where fault detection filters were developed for systems subjected to communication delays and missing data. Anomaly detection over networks is another important topic, since anomalies may affect the performance of network control systems. In [67], a sliding mode observer was devised for anomaly detection over Transmission Control Protocol (TCP) networks.

Distributed systems. Distributed systems are based on interconnection of many subsystems. These subsystems have access to local measurements and local controllers. With respect to networked systems, here the communication is limited to the neighbors of each local module. The advantages with respect to networked systems are a lower use of network resources, cost-effectiveness, and convenience for expansion. However, due to this constraints on resources and communication, real-time monitoring is more difficult [79]. The basic idea is to design local estimators or fault detection filters, and then employ a consensus strategy to ensure the performance of the whole network [48].

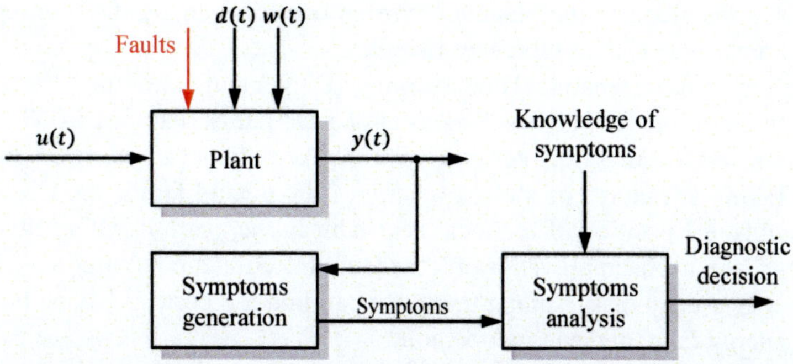

Fig. 3.12 Schematic description of the signal-processing-based scheme. Adapted by permission from Springer Nature. Model-Based Fault Diagnosis Techniques Design Schemes, Algorithms and Tools by Steven X. Ding © (2013)

3.2.2 Signal-Based Approach

For large-scale or complex processes, it is not always possible to develop an effective model of the system. Signal-based methods utilize measured signals rather than explicit input–output models for fault diagnosis. The basic assumption is that faults in the system affect the measured output signals. By extracting features, or symptoms, from the measurements, a diagnostic decision is made by comparing the actual patterns' values with prior knowledge on the symptoms of the healthy system. Signal-based approaches are especially useful when the monitored process is characterized by an oscillating or cyclic time behavior, such as in the monitoring of bearings. A schematic of the method is depicted in Fig. 3.12.

The features to be extracted for symptom (or pattern) analysis can be

- *Time domain*: such as mean, trends, standard deviation, phases, slope, and magnitudes such as peak and root mean square;
- *Frequency domain*: such as spectrum analysis;
- *Time–Frequency domain*: such as Short-Time-Fourier Transform (STFT) and wavelet analysis.

3.2.2.1 Time-Domain Signal-Based Methods

Extracting features from the time-domain behavior of measured signals is perhaps the most intuitive approach for signal-based fault diagnosis. For instance, much work has been performed on monitoring electrical motors on the basis of electrical measurements such as currents and voltages. In particular, the employed features entail the computation of the root mean square value of the current in transistors, the derivative of the Park's vector phase angle for Permanent-Magnet Synchronous Machines

(PMSMs), the slope of the induction current over time in DC-DC converters, and many more: see [31] and references therein.

The use of fast Dynamic Time Warping (DTW) and Correlated Kurtosis (CK) has been employed in [42] for diagnosis of gear faults. The fast DTW algorithm was employed to extract the periodic impulse excitations caused by the faulty gear tooth. Taking advantage of the periodicity of the geared faults, the CK algorithm can identify the position of the local gear fault in the gearbox. Other time-domain features that have been proved useful for fault detection are reported in [51], with a focus on gear diagnosis. Among these, we can find the kurtosis, shape factor, crest factor, energy ratio, and energy operator.

An alternative approach is to extract features in a two-dimensional domain: in [15] measured vibration signals were translated into images, and image processing techniques such as Scale-Invariant Feature Transform (SIFT) were employed to detect symptoms.

3.2.2.2 Frequency-Domain Signal-Based Methods

Frequency-domain signal-based methods employ Discrete Fourier Transform (DFT) to compute the spectrum of a signal. A typical example is the Motor Current Signature Analysis (MCSA), which uses the spectral analysis of the stator current to sense rotor faults associated with broken rotor bars and mechanical balance [59].

The analysis of vibration signals is a common method for condition monitoring and diagnosis for mechanical equipment such as gearbox and bearings [69]. In [51], various frequency-domain features are reported for gear monitoring, such as mean frequency, frequency center, root mean square frequency, and standard deviation frequency.

3.2.2.3 Time-Frequency-Domain Signal-Based Methods

Time–frequency signal processing techniques are necessary in all the cases where the measured signals undergo transient dynamic conditions such as unbalanced supply voltages, varying load, or load torque oscillations. Time-frequency analysis can evaluate the frequency content with respect to different time instants. The most used approaches are the Short-Time Fourier Transform (STFT), Wavelet Transforms (WT) [39], Hilbert–Huang Transform (HHT) [91], and Wigner–Ville Distribution (WVD) [16]. Again, see [51] for a variety of time–frequency features for gear diagnosis. A notable mention regards the Spectral Kurtosis (SK) algorithm [70], widely employed as a preprocessing step for envelope analysis in rolling bearings diagnosis. The spectral kurtosis method aims to find the frequency band of the signal where "most impulsive" events occur. This band will then be used to band-pass filter the signal (e.g., an accelerometer measure) in order to enhance the signal-to-noise ratio. The envelope of the filtered signal is then computed and, subsequently, the Fourier

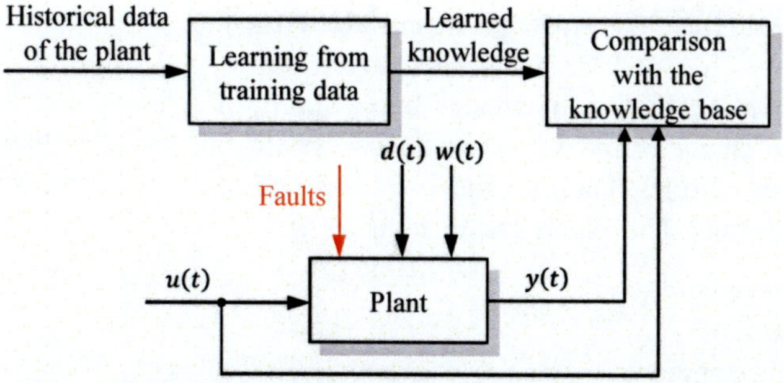

Fig. 3.13 Schematic description of the knowledge-based fault diagnosis scheme

transform of the envelope is evaluated in search of symptoms (e.g., the appearance specific fault frequencies in the measured signal).

3.2.3 Knowledge-Based Approach

Model-based and signal-based methods require the presence of *a priori* (structured) knowledge. In fact, they need a known model or specific signal symptoms to be looked for, respectively. Knowledge-based fault diagnosis methods do not have this prior information about system behavior or symptoms, but they assume that this *indication is hidden inside the measured data*, and just need to be discovered. Thus, knowledge-based fault diagnosis methods are based on a *large volume of historical data*. By using a variety of artificial intelligence techniques (either symbolic intelligence or computing intelligence), the prior (unstructured) knowledge that lies in the historical data can be leveraged and use to build a *knowledge base*. Then, real-time data are checked for consistency with the learned knowledge base, and a diagnostic decision is made by employing a classifier (that can be implemented with machine learning models, via simple thresholds or via an expert system with logical rules). Figure 3.13 depicts a schematic of a knowledge-based fault diagnosis system. These methods are often referred also as *data-driven fault diagnosis* approaches (not to be confused with those reported in Sect. 3.2.1.3).

The creation of a knowledge base can be performed via two different modalities:

- qualitative knowledge-based methods;
- quantitative knowledge-based methods.

Therefore, knowledge-based fault diagnosis methods can be classified into qualitative methods and quantitative methods.

3.2.3.1 Qualitative Knowledge-Based Methods

The most used qualitative knowledge-based approach is the *expert systems*. Expert systems are characterized by a rule-based system that encodes human expertise through a set of rules. Fault diagnosis is then realized by running well-developed search algorithms. The expert system consists of

- a knowledge base;
- a database;
- an inference engine;
- an explanation component.

With expert systems, the learning phase is executed by the human, who can insert his experience into the set of rules. Expert systems provide a transparent explanation of their decision to the user. However, they tend to be system specific, having low generality and low expandability. A recent work was proposed in [5] in order to make expert systems more general and flexible.

In many practical industrial processes, such as in the chemical sector, process malfunctions leave a distinct trend in the sensors monitored. *Qualitative Trend Analysis* (QTA) is another knowledge-based method that can be used to associate process trends from noisy data to associate fault trends in the database [87].

3.2.3.2 Quantitative Knowledge-Based Methods

Quantitative knowledge-based fault diagnosis entails the extraction of information (or features) from data, leading to a pattern recognition problem. The features can be computed by using statistical or nonstatistical techniques.

Statistical analysis knowledge-based fault diagnosis. Among the most famous statistical analysis techniques, there are Principal Component Analysis (PCA), Partial Least Squares (PLS), Canonical Variate Analysis (CVA), and Independent Component Analysis (ICA).

Principal component analysis is the most popular statistically based monitoring technique, which is utilized to find factors with a much lower dimension than the original dataset. PCA is able to generate statistics describing the variations of data, and of the noise acting on the data. Typically, the Hotelling's T^2 statistic and the Q-statistic, also known as the Squared Prediction Error (SPE), are used for the detection of an abnormal (out-of-control) situation [66]. Predefined thresholds, computed-based on statistical distributions, are available for these indexes, in order to accomplish fault detection [71].

While the PCA-like process monitoring can detect and diagnose abnormal situations in the process data, it cannot tell if the detected abnormal situation will lead to product quality problems. *Partial Least Squares*, also known as Projection to Latent Structures, is a dimensionality reduction technique for maximizing the covariance between two datasets: a predictor and a predicted one [71]. A popular application of

PLS is to select the predicted set to contain product quality data, and the predictors set to contain all other process variables. Such inferential models (also known as *soft sensors*) can be used for the online forecast of product quality data. Well-written surveys on PCA, PLS, and their extensions are available in [66, 92].

Canonical Variate Analysis is a dimensionality reduction technique from multivariate statistical analysis, optimal in terms of maximizing a correlation measure between two sets of variables [71]. This relates CVA to PLS. CVA has been furthermore employed in the context of subspace system identification to estimate state-space linear dynamic models from data. The identified system states can be employed to generate monitoring statistics similar to those of PCA [89].

Independent Component Analysis plays an important role in practical industrial processes since it allows latent variables not to follow a Gaussian distribution. A recent fault isolation method was proposed in [95] for non-Gaussian nonlinear processes.

Non-statistical analysis knowledge-based fault diagnosis. *Machine learning* methods, both supervised and unsupervised, can be cast in the framework of nonstatistical analysis knowledge-based fault diagnosis. Here the aim is to define a set of system conditions to be monitored. Each condition can be, apart from the healthy state, a specific fault type. Then, a machine learning model is trained on data that originated from each condition. During real-time monitoring, the model is used to classify the features extracted from online data into one of the former classes, see e.g., [82].

Fuzzy Logic (FL) partitions a feature space into fuzzy sets and uses fuzzy rules for reasoning, providing an approximate human reasoning scheme. FL can be employed successfully for fault diagnosis. As an example, in [97], FL was employed to represent a fuzzy knowledge base that was extracted from the analysis of currents data, and applied to detect misfiring in the switches in a Pulse Width Modulation (PWM) source inverter induction motor drive.

3.2.4 Hybrid Approach

A hybrid approach is often exploited in order to combine the advantages of each of the different fault diagnosis methodologies. Specifically, model-based methods are able to cope with the dynamic constraints of the process, diagnosing unknown faults with a small amount of real-time data. However, they require an explicit model describing the input/output relationship, and this can be a drawback for very complex and articulated systems. Signal-based and knowledge-based methods can be employed, where a process model is too complex and difficult to obtain.

The signal-based method generally uses output data in order to compute the symptoms for fault diagnosis. Therefore, it pays less attention on system dynamic inputs. The diagnosis performance can be compromised in presence of unknown input disturbances and varying operative conditions.

The knowledge-based method may suffer from high computational costs and may not work well for identifying unknown fault types (i.e., a condition not present in the knowledge base).

As a relevant example, in [75] a hybrid knowledge-based and model-based method is proposed for the FDI of chemical reactors with high nonlinearities and complex dynamics.

3.2.5 *Active Approach*

Active fault diagnosis consists of purposely inject a control input to the process, under a test time interval. The aim is to enhance the detectability of faulty modes of the system. In this case, the additional input should not compromise the system performance. A unified formulation for active fault detection and control is proposed [76]. A recent application of active fault diagnosis methods can be found in [53] for monitoring stator windings and rotor permanent magnets of PMSMs.

3.3 State-of-the-Art of Monitoring Approaches for Airborne Electro-Mechanical Actuators and Systems

Fault diagnosis in aerospace systems. The presented monitoring approaches have found many examples in the aerospace environment, especially for actuators and sensors. The use of multiple hardware redundancy is used in digital Fly-By-Wire flight control systems, e.g., the AIRBUS 320 and its derivatives [28]. General guidelines of AIRBUS FDI and FTC practices are given in [37]. The main problematics for condition monitoring of EMAs in aerospace have been reviewed in [81].

Model-based methods using observers have been applied to aircraft sensors in [63]. A recent result on fault isolation for aircraft engines is given in [94]. Unknown input observers were applied to aircraft actuators in [88]. The use of a particle filter for fault detection and isolation for EMAs in aerospace is proposed in [56]. An example of modeling and simulation of an aerospace EMAs is given in [30]. In [7], the authors developed a model-based approach to prognostics (i.e., predicting the residual life of the system/component) and health management, for actuator fault detection and failure progression. The parity space approach was used to detect faults in an aerospace actuator in [60]. The authors in [17] proposed a health-monitoring method for EMAs based on position predictive models. A model-based prognostic method for the free-play identification in flight EMAs has been devised in [19]. Faults in aerospace EMAs for unmanned aerial system flight controls were faced in [18]. Authors in [62] employed model-based methods to diagnose loss of efficiency faults in flight EMAs, while [41] devised a model-based approach to tackle elevator runaway and elevator jamming faults.

In [72], a statistical method for the detection of sensor abrupt faults in aircraft control systems was presented, where the covariance of the sensing signals was used for feature extraction. A signal-based method based on statistical change detection is presented in [57, 58]. Vibration signatures for high-criticality jam and a low-criticality spall (metal flaking) in the actuator ball screw mechanism in flight EMAs have been considered in [47].

The authors in [65] defined the structure that a knowledge-based approach for aerospace condition monitoring, using expert systems, should have. A comparison of supervised machine learning knowledge-based method for fault detection and isolation in aerospace EMA is proposed in [55]. For an unsupervised approach based on clustering, see [54].

A combined model-based and quantitative knowledge-based prognostic health management software for aerospace EMA is presented in [2]. A hybrid method, combining signal-based and knowledge-based method, based on machine learning classifiers was proposed in [14].

Fault diagnosis references and books. Model-based fault diagnosis is well presented in the books [11, 22, 44, 77]. The work [45] contains many practical applications of fault diagnosis, including a BLDC motor for an aircraft cabin pressure valve. A MATLAB toolbox for the book [84] is available in [83]. The book [84] contains practical examples of fault diagnosis for flight actuators and sensors. A collection of fault diagnosis approaches on flight EMAs and systems is thoroughly presented in [6, 98], while [3] focuses on sensors faults for aircraft. Signal-based methods for the analysis of vibration signals are described in [69]. A good book on statistical knowledge-based methods for process data is [71]. These methods, along with the data-driven design of analytical residual generators, are presented in [23]. Fault-tolerant control and reconfiguration is faced in the well-written book of [4]. A comprehensive review and comparison of the different methodologies have been presented in [80]. For a three-part survey of methodologies in the field of process fault diagnosis, see [85–87].

3.4 Summary

In the first part of this chapter, the basic definitions and terminology used in the fault diagnosis community are given, aiming to harmonize and integrate the concepts and the nomenclatures coming from different research fields. In the second part of the chapter, the main approaches to fault diagnosis have been classified and presented. Finally, current applications of fault diagnosis and condition monitoring for EMA in aerospace have been reviewed.

References

1. Alwi H, Edwards C (2014) Robust fault reconstruction for linear parameter varying systems using sliding mode observers. Int J Robust Nonlinear Control 24(14):1947–1968. https://doi.org/10.1002/rnc.3009
2. Balaban E, Saxena A, Narasimhan S, Roychoudhury I, Goebel K (2011) Experimental validation of a prognostic health management system for electro-mechanical actuators. In: Infotech@ aerospace, p 1518 (2011). https://doi.org/10.2514/6.2011-1518
3. Benini M, Castaldi P, Simani S et al (2009) Fault diagnosis for aircraft system models: an introduction from fault detection to fault tolerance. VDM Verlag Dr. Muller Aktiengesellschaft & Co, KG
4. Blanke M, Kinnaert M, Lunze J, Staroswiecki M, Schröder J (2016) Diagnosis and fault-tolerant control. Springer, Berlin, Heidelberg (2016). https://doi.org/10.1007/978-3-662-47943-8
5. Bo M, Zhi-nong J, Zhong-qing W (2012) Development of the task-based expert system for machine fault diagnosis. J Phys: Conf Ser 364:012043. IOP Publishing
6. Brandt A (2011) Noise and vibration analysis: signal analysis and experimental procedures. Wiley. https://doi.org/10.1002/9780470978160
7. Byington CS, Watson M, Edwards D, Stoelting P (2004) A model-based approach to prognostics and health management for flight control actuators. In: 2004 IEEE Aerospace Conference Proceedings (IEEE Cat. No.04TH8720), vol 6, pp 3551–3562 (2004). https://doi.org/10.1109/AERO.2004.1368172
8. Cabasino MP, Giua A, Seatzu C (2010) Fault detection for discrete event systems using petri nets with unobservable transitions. Automatica 46(9):1531–1539. https://doi.org/10.1016/j.automatica.2010.06.013
9. Chen H, Lu S (2013) Fault diagnosis digital method for power transistors in power converters of switched reluctance motors. IEEE Trans Indust Electron 60(2):749–763. https://doi.org/10.1109/TIE.2012.2207661
10. Chen J, Patton RJ (1971) Failure accommodation in linear system through self reorganization. PhD dissertation, MIT, Cambridge, MA, USA
11. Chen J, Patton RJ (1999) Robust model-based fault diagnosis for dynamic systems. Springer US. https://doi.org/10.1007/978-1-4615-5149-2
12. Chen J, Zhang H (1991) Robust detection of faulty actuators via unknown input observers. Int J Syst Sci 22(10):1829–1839. https://doi.org/10.1080/00207729108910753
13. Chen YQ, Fink O, Sansavini G (2018) Combined fault location and classification for power transmission lines fault diagnosis with integrated feature extraction. IEEE Tran Industr Electron 65(1):561–569. https://doi.org/10.1109/TIE.2017.2721922
14. Chirico AJ, Kolodziej JR (2014) A data-driven methodology for fault detection in electromechanical actuators. J Dyn Syst Measur Control **136**(4), 041,025 (2014). https://doi.org/10.1115/1.4026835
15. Chong UP, Do V (2011) Signal model-based fault detection and diagnosis for induction motors using features of vibration signal in two-dimension domain. J Mechan Eng 57(9):655–666. https://doi.org/10.5545/sv-jme.2010.162
16. Climente-Alarcon V, Antonino-Daviu JA, Riera-Guasp M, Vlcek M (2014) Induction motor diagnosis by advanced notch fir filters and the wignerville distribution. IEEE Trans Industr Electron 61(8):4217–4227. https://doi.org/10.1109/TIE.2013.2286581
17. Di Rito G, Schettini F (2018) Health monitoring of electromechanical flight actuators via position-tracking predictive models. Adv Mech Eng **10**(4) (2018). https://doi.org/10.1177/1687814018768146
18. Di Rito G, Schettini F, Galatolo R (2017) Model-based health-monitoring of an electromechanical actuator for unmanned aerial system flight controls. In: 2017 IEEE international workshop on metrology for aerospace (MetroAeroSpace), pp 502–511 (2017). https://doi.org/10.1109/MetroAeroSpace.2017.7999626

19. Di Rito G, Schettini F, Galatolo R (2018) Model-based prognostic health-management algorithms for the freeplay identification in electromechanical flight control actuators. In: 2018 5th IEEE international workshop on metrology for AeroSpace (MetroAeroSpace), pp 340–345. IEEE (2018). https://doi.org/10.1109/MetroAeroSpace.2018.8453552
20. Ding S (2014) Data-driven design of monitoring and diagnosis systems for dynamic processes: a review of subspace technique based schemes and some recent results. J Process Control 24(2):431–449. https://doi.org/10.1016/j.jprocont.2013.08.011. ADCHEM 2012 Special Issue
21. Ding S, Zhang P, Naik A, Ding E, Huang B (2009) Subspace method aided data-driven design of fault detection and isolation systems. J Process Control 19(9):1496–1510. https://doi.org/10.1016/j.jprocont.2009.07.005
22. Ding SX (2013) Model-based fault diagnosis techniques: design schemes, algorithms, and tools, 2nd edn. Springer Publishing Company, Incorporated
23. Ding SX (2014) Data-driven design of fault diagnosis and fault-tolerant control systems, 1st edn. Springer, London
24. Ding X, Frank PM (1990) Fault detection via factorization approach. Syst Control Lett 14(5):431–436. https://doi.org/10.1016/0167-6911(90)90094-B
25. Dong J (2009) Data driven fault tolerant control: a subspace approach. PhD thesis, Technische Universiteit Delft
26. Dong J, Verhaegen M (2009) Subspace based fault detection and identification for lti systems. 7th IFAC symposium on fault detection, supervision and safety of technical processes 42(8):330–335 (2009). https://doi.org/10.3182/20090630-4-ES-2003.00055
27. Fadda G, Pilloni A, Pisano A, Usai E, Marjanović A, Vujnović S (2015) Sensor fault diagnosis in water-steam power plant: A combined observer-based/pattern-recognition approach. In: Recent advances in sliding modes (RASM), 2015 international workshop on, pp 1–7. IEEE (2015). https://doi.org/10.1109/RASM.2015.7154643
28. Favre C (1994) Fly-By-Wire for commercial aircraft: the airbus experience. Int J Control 59(1):139–157. https://doi.org/10.1080/00207179408923072
29. Foo GHB, Zhang X, Vilathgamuwa DM (2013) A sensor fault detection and isolation method in interior permanent-magnet synchronous motor drives based on an extended kalman filter. IEEE Trans Industr Electron 60(8):3485–3495. https://doi.org/10.1109/TIE.2013.2244537
30. Fu J, Maré JC, Fu Y (2017) Modelling and simulation of flight control electromechanical actuators with special focus on model architecting, multidisciplinary effects and power flows. Chinese J Aeronaut 30(1):47–65. https://doi.org/10.1016/j.cja.2016.07.006
31. Gao Z, Cecati C, Ding SX (2015) A survey of fault diagnosis and fault-tolerant techniques - part i: Fault diagnosis with model-based and signal-based approaches. IEEE Trans Industr Electron 62(6):3757–3767. https://doi.org/10.1109/TIE.2015.2417501
32. Gao Z, Cecati C, Ding SX (2015) A survey of fault diagnosis and fault-tolerant techniques—Part II: Fault diagnosis with knowledge-based and hybrid/active approaches. IEEE Trans Industr Electron 62(6):3768–3774. https://doi.org/10.1109/TIE.2015.2419013
33. Gao Z, Wang H (2006) Descriptor observer approaches for multivariable systems with measurement noises and application in fault detection and diagnosis. Syst Control Lett 55(4):304–313. https://doi.org/10.1016/j.sysconle.2005.08.004
34. Gertler J (1998) Fault detection and diagnosis in engineering systems. CRC Press
35. Gertler JJ (1988) Survey of model-based failure detection and isolation in complex plants. IEEE Control Syst Mag 8(6):3–11. https://doi.org/10.1109/37.9163
36. Gholizadeh M, Salmasi FR (2014) Estimation of state of charge, unknown nonlinearities, and state of health of a lithium-ion battery based on a comprehensive unobservable model. IEEE Trans Industr Electron 61(3):1335–1344. https://doi.org/10.1109/TIE.2013.2259779
37. Goupil P (2011) Airbus state of the art and practices on fdi and ftc in flight control system. Control Eng Pract 19(6):524–539. https://doi.org/10.1016/j.conengprac.2010.12.009. SAFEPROCESS 2009
38. Grastien A, Anbulagan A (2013) Diagnosis of discrete event systems using satisfiability algorithms: a theoretical and empirical study. IEEE Trans Autom Control 58(12):3070–3083. https://doi.org/10.1109/TAC.2013.2275892

39. Gritli Y, Zarri L, Rossi C, Filippetti F, Capolino G, Casadei D (2013) Advanced diagnosis of electrical faults in wound-rotor induction machines. IEEE Trans Industr Electron 60(9):4012–4024. https://doi.org/10.1109/TIE.2012.2236992
40. He X, Wang Z, Zhou D (2009) Robust fault detection for networked systems with communication delay and data missing. Automatica 45(11):2634–2639. https://doi.org/10.1016/j.automatica.2009.07.020
41. Hecker S, Varga A, Ossmann D (2011) Diagnosis of actuator faults using lpv-gain scheduling techniques. In: AIAA guidance, navigation, and control conference, p 6680
42. Hong L, Dhupia JS (2014) A time domain approach to diagnose gearbox fault based on measured vibration signals. J Sound Vibr 333(7):2164–2180. https://doi.org/10.1016/j.jsv.2013.11.033
43. IEC 61508 - Functional safety of electrical/electronic/programmable electronic systems (1997)
44. Isermann R (2006) Fault-Diagnosis systems—an introduction from fault detection to fault tolerance. Springer, Berlin, Heidelberg. https://doi.org/10.1007/3-540-30368-5
45. Isermann R (2011) Fault-diagnosis applications: model-based condition monitoring: actuators, drives, machinery, plants, sensors, and fault-tolerant systems. Springer Sc Bus Media. https://doi.org/10.1007/978-3-642-12767-0
46. Isermann R, Ballé P (1997) Trends in the application of model-based fault detection and diagnosis of technical processes. Control Eng Pract 5(5):709–719. https://doi.org/10.1016/S0967-0661(97)00053-1
47. Ismail MAA, Balaban E, Spangenberg H (2016) Fault detection and classification for flight control electromechanical actuators. In: 2016 IEEE aerospace conference, pp 1–10 (2016). https://doi.org/10.1109/AERO.2016.7500784
48. Keliris C, Polycarpou MM, Parisini T (2013) A distributed fault detection filtering approach for a class of interconnected continuous-time nonlinear systems. IEEE Trans Autom Control 58(8):2032–2047. https://doi.org/10.1109/TAC.2013.2253231
49. Koenig D (2005) Unknown input proportional multiple-integral observer design for linear descriptor systems: application to state and fault estimation. IEEE Trans Autom Control 50(2):212–217. https://doi.org/10.1109/TAC.2004.841889
50. Kulesza Z, Sawicki JT, Gyekenyesi AL (2013) Robust fault detection filter using linear matrix inequalities' approach for shaft crack diagnosis. J Vibr Control 19(9):1421–1440. https://doi.org/10.1177/1077546312447838
51. Lei Y, Zuo MJ, He Z, Zi Y (2010) A multidimensional hybrid intelligent method for gear fault diagnosis. Expert Syst Appl 37(2):1419–1430. https://doi.org/10.1016/j.eswa.2009.06.060
52. Levy R, Arogeti SA, Wang D (2014) An integrated approach to mode tracking and diagnosis of hybrid systems. IEEE Trans Industr Electron 61(4):2024–2040. https://doi.org/10.1109/TIE.2013.2262761
53. Liu K, Zhu ZQ, Stone DA (2013) Parameter estimation for condition monitoring of pmsm stator winding and rotor permanent magnets. IEEE Trans Industr Electron 60(12):5902–5913. https://doi.org/10.1109/TIE.2013.2238874
54. Mazzoleni M, Formentin S, Previdi F, Savaresi SM (2014) Fault detection via modified principal direction divisive partitioning and application to aerospace electro-mechanical actuators. In: Decision and control (CDC), 2014 IEEE 53rd annual conference on, pp 5770–5775. IEEE (2014). https://doi.org/10.1109/CDC.2014.7040292
55. Mazzoleni M. Maccarana Y, Previdi F (2017) A comparison of data-driven fault detection methods with application to aerospace electro-mechanical actuators. 20th IFAC World Congress **50**(1):12,797–12,802 (2017). https://doi.org/10.1016/j.ifacol.2017.08.1837
56. Mazzoleni M, Maroni G, Maccarana Y, Formentin S, Previdi F (2017) Fault detection in airliner electro-mechanical actuators via hybrid particle filtering. 20th IFAC World Congress **50**(1), 2860–2865. https://doi.org/10.1016/j.ifacol.2017.08.640
57. Mazzoleni M, Scandella M. Maccarana Y, Previdi F, Pispola G, Porzi N (2018) Condition assessment of electro-mechanical actuators for aerospace using relative density-ratio estimation. 18th IFAC symposium on system identification (SYSID) 51(15):957–962 (2018). https://doi.org/10.1016/j.ifacol.2018.09.070

58. Mazzoleni M, Scandella M, Maccarana Y, Previdi F, Pispola G, Porzi N (2018) Condition monitoring of electro-mechanical actuators for aerospace using batch change detection algorithms. In: 2018 IEEE conference on control technology and applications (CCTA), pp 1747–1752. IEEE. https://doi.org/10.1109/CCTA.2018.8511334
59. Nandi S, Toliyat HA, Li X (2005) Condition monitoring and fault diagnosis of electrical motors–a review. IEEE Trans Energy Conver 20(4):719–729. https://doi.org/10.1109/TEC.2005.847955
60. Odendaal HM, Jones T (2014) Actuator fault detection and isolation: an optimised parity space approach. Control Eng Practice 26:222–232. https://doi.org/10.1016/j.conengprac.2014.01.013
61. Orchard ME, Vachtsevanos GJ (2009) A particle-filtering approach for on-line fault diagnosis and failure prognosis. Trans Inst Measur Control 31(3–4):221–246. https://doi.org/10.1177/0142331208092026
62. Ossmann D, Varga A (2015) Detection and identification of loss of efficiency faults of flight actuators. Int J Appl Math Comput Sci 25(1):53–63. https://doi.org/10.1515/amcs-2015-0004
63. Patton RJ, Willcox SW, Winter JS (1987) Parameter-insensitive technique for aircraft sensor fault analysis. J Guidance Control Dyn 10(4):359–367. https://doi.org/10.2514/3.20226
64. Pencolé Y, Cordier MO (2005) A formal framework for the decentralised diagnosis of large scale discrete event systems and its application to telecommunication networks. Artif Intell 164(1–2):121–170. https://doi.org/10.1016/j.artint.2005.01.002
65. Phillips P, Diston D (2011) A knowledge driven approach to aerospace condition monitoring. Know Based Syst 24(6):915–927. https://doi.org/10.1016/j.knosys.2011.04.008
66. Qin SJ (2012) Survey on data-driven industrial process monitoring and diagnosis. Ann Rev Control 36(2):220–234. https://doi.org/10.1016/j.arcontrol.2012.09.004
67. Rahmé S, Labit Y, Gouaisbaut F, Floquet T (2013) Sliding modes for anomaly observation in tcp networks: from theory to practice. IEEE Trans Control Syst Technol 21(3):1031–1038. https://doi.org/10.1109/TCST.2012.2198648
68. Raich A, Cinar A (1996) Statistical process monitoring and disturbance diagnosis in multivariable continuous processes. AIChE J 42(4):995–1009. https://doi.org/10.1002/aic.690420412
69. Randall RB (2011) Vibration-based condition monitoring: industrial, aerospace and automotive applications. Wiley
70. Randall RB, Antoni J (2011) Rolling element bearing diagnostics–a tutorial. Mech Syst Signal Process 25(2):485–520. https://doi.org/10.1016/j.ymssp.2010.07.017
71. Russell EL, Chiang LH, Braatz RD (2012) Data-driven methods for fault detection and diagnosis in chemical processes. Springer Sci Bus Media. https://doi.org/10.1007/978-1-4471-0409-4
72. Samara PA, Fouskitakis GN, Sakellariou JS, Fassois SD (2008) A statistical method for the detection of sensor abrupt faults in aircraft control systems. IEEE Trans Control Syst Technol 16(4):789–798. https://doi.org/10.1109/TCST.2007.903109
73. van Schrick D (1997) Remarks on terminology in the field of supervision, fault detection and diagnosis. IFAC Symposium on Fault Detection, Supervision and Safety for Technical Processes (SAFEPROCESS 97), Kingston upon Hull, UK, 26-28 August 1997 30(18):959–964. https://doi.org/10.1016/S1474-6670(17)42524-9
74. Schumann A, Pencolé Y, Thiébaux S, et al (2007) A spectrum of symbolic on-line diagnosis approaches. In: Proceeedings of the national conference on artificial intelligence, vol 22, p 335. Menlo Park, CA; Cambridge, MA; London; AAAI Press; MIT Press; 1999
75. Sheibat-Othman N, Laouti N, Valour JP, Othman S (2014) Support vector machines combined to observers for fault diagnosis in chemical reactors. The Canadian J Chem Eng 92(4):685–695. https://doi.org/10.1002/cjce.21881
76. Simandl M, Puncochar I (2009) Active fault detection and control: unified formulation and optimal design. Automatica 45(9):2052–2059 (2009). https://doi.org/10.1016/j.automatica.2009.04.028. http://www.sciencedirect.com/science/article/pii/S0005109809002210
77. Simani S, Fantuzzi C, Patton JR (2003) Model-based fault diagnosis in dynamic systems using identification techniques. Springer, London . https://doi.org/10.1007/978-1-4471-3829-7

78. Simani S, Farsoni S (2018) Fault diagnosis and sustainable control of wind turbines. Butterworth-Heinemann. https://doi.org/10.1016/C2016-0-04286-9
79. Stankovic SS, Stankovic MS, Stipanovic DM (2009) Consensus based overlapping decentralized estimation with missing observations and communication faults. Automatica 45(6):1397–1406. https://doi.org/10.1016/j.automatica.2009.02.014
80. Tidriri K, Chatti N, Verron S, Tiplica T (2016) Bridging data-driven and model-based approaches for process fault diagnosis and health monitoring: a review of researches and future challenges. Ann Rev Control 42:63–81. https://doi.org/10.1016/j.arcontrol.2016.09.008
81. Todeschi M, Baxerres L (2015) Health monitoring for the flight control emas. IFAC-PapersOnLine 48(21):186–193. https://doi.org/10.1016/j.ifacol.2015.09.526. 9th IFAC Symposium on Fault Detection, Supervision andSafety for Technical Processes SAFEPROCESS 2015
82. Toma S, Capocchi L, Capolino G (2013) Wound-rotor induction generator inter-turn short-circuits diagnosis using a new digital neural network. IEEE Trans Indust Electron 60(9):4043–4052. https://doi.org/10.1109/TIE.2012.2229675
83. Varga A, FDITOOLS—The Fault Detection and Isolation Tools for MATLAB. https://sites.google.com/site/andreasvargacontact/home/software/fditools
84. Varga A (2017) Solving fault diagnosis problems, vol 84. Springer International Publishing. https://doi.org/10.1007/978-3-319-51559-5
85. Venkatasubramanian V (2003) A review of process fault detection and diagnosis part i: quantitative model-based methods. Comput Chem Eng 27(3):293–311. https://doi.org/10.1016/S0098-1354(02)00160-6
86. Venkatasubramanian V, Rengaswamy R, Kavuri SN (2003) A review of process fault detection and diagnosis part ii: qualitative models and search strategies. Comput Chem Eng 27(313Á/326). https://doi.org/10.1016/S0098-1354(02)00161-8
87. Venkatasubramanian V, Rengaswamy R, Kavuri SN, Yin K (2003) A review of process fault detection and diagnosis: Part iii: process history based methods. Comput Chem Engi 27(3):327–346. https://doi.org/10.1016/S0098-1354(02)00162-X
88. Wang D, Lum KY (2007) Adaptive unknown input observer approach for aircraft actuator fault detection and isolation. Int J Adapt Control Signal Process 21(1):31–48. https://doi.org/10.1002/acs.936
89. Wang Y, Seborg DE, Larimore WE (1997) Process monitoring using canonical variate analysis and principal component analysis. IFAC symposium on advanced control of chemical processes 1997 (ADCHEM '97), Banff, Canada, 9–11 June 30(9):577–582. https://doi.org/10.1016/S1474-6670(17)43211-3
90. Willsky AS (1976) A survey of design methods for failure detection in dynamic systems. Automatica 12(6):601–611. https://doi.org/10.1016/0005-1098(76)90041-8
91. Yan R, Gao RX (2006) Hilbert-huang transform-based vibration signal analysis for machine health monitoring. IEEE Trans Instrum Measur 55(6):2320–2329. https://doi.org/10.1109/TIM.2006.887042
92. Yin S, Ding SX, Xie X, Luo H (2014) A review on basic data-driven approaches for industrial process monitoring. IEEE Trans Industr Electron 61(11):6418–6428. https://doi.org/10.1109/TIE.2014.2301773
93. Zhang K, Jiang B, Cocquempot V, Zhang H (2013) A framework of robust fault estimation observer design for continuous-time/discrete-time systems. Optimal Control Appl Methods 34(4):442–457. https://doi.org/10.1002/oca.2031
94. Zhang X, Tang L, Decastro J (2013) Robust fault diagnosis of aircraft engines: a nonlinear adaptive estimation-based approach. IEEE Trans Control Syst Technol 21(3):861–868. https://doi.org/10.1109/TCST.2012.2187057
95. Zhang Y, Yang N, Li S (2014) Fault isolation of nonlinear processes based on fault directions and features. IEEE Trans Control Syst Technol 22(4):1567–1572. https://doi.org/10.1109/TCST.2013.2283925
96. Zhao F, Koutsoukos X, Haussecker H, Reich J, Cheung P (2005) Monitoring and fault diagnosis of hybrid systems. IEEE Trans Syst Man Cybern Part B (Cybernetics) 35(6):1225–1240 (2005). https://doi.org/10.1109/TSMCB.2005.850178

97. Zidani F, Diallo D, Benbouzid MEH, Nait-Said R (2008) A fuzzy-based approach for the diagnosis of fault modes in a voltage-fed pwm inverter induction motor drive. IEEE Trans Industr Electron 55(2):586–593. https://doi.org/10.1109/TIE.2007.911951
98. Zolghadri A, Henry D, Cieslak J, Efimov D, Goupil P (2014) Fault diagnosis and fault-tolerant control and guidance for aerospace vehicles. Springer

Chapter 4
Fault Diagnosis and Condition Monitoring of Aircraft Electro-Mechanical Actuators

Outline of the chapter. This chapter is structured as follows.[1] Section 4.1 motivates the challenges of designing and developing analytical fault diagnosis methods for airborne EMAs.

Section 4.2 describes experimental projects, carried out by different funding sources, related to the development of reliable electro-mechanical actuators for aerospace. This includes both the design of software fault diagnosis methods and the development of innovative hardware configurations.

The subsequent sections will present different fault diagnosis and condition monitoring approaches, mainly applied to the previously introduced project cases. The monitoring methodologies are presented starting from problems where more information on physics-of-failure are available (model-based approaches), to problems where these information are poor (signal-based approaches), and concluding with problems where no prior information about the faults is given (knowledge-based approaches).

Following this rationale, see Fig. 4.1, model-based approaches are presented firstly in Sect. 4.3 with examples of fault detection and fault prognosis strategies. A review of signal-based methods is then presented in Sect. 4.4. Lastly, examples of knowledge-based approaches are described in Sect. 4.5.

The relation between diagnosis methods and related projects is described herein.

FP7 HOLMES project. Section 4.2.1 introduces the European FP7 HOLMES project, focused on *fault detection and isolation* of secondary flight control actuators. The tested approaches include:

- *Knowledge-based*: supervised machine learning method, Sect. 4.5.1.

H2020 REPRISE project. Sections 4.2.2 and 4.2.3 highlight, respectively, the first and second phases of the European H2020 REPRISE project, focused on *condition monitoring* of primary flight control actuators. The tested approaches include:

---

[1]Sections 4.2.2 and 4.5.3 are based on the work [40]—originally published open access and licensed under CC-BY 4.0. https://ieeexplore.ieee.org/stamp/stamp.jsp?tp=&arnumber=8878102. Sections 4.2.4 and 4.3.1 are based on the work [16]—originally published open access and licensed under CC-BY 4.0. https://journals.sagepub.com/doi/pdf/10.1177/1687814018768146.

© Springer Nature Switzerland AG 2021
M. Mazzoleni et al., *Electro-Mechanical Actuators for the More Electric Aircraft*, Advances in Industrial Control, https://doi.org/10.1007/978-3-030-61799-8_4

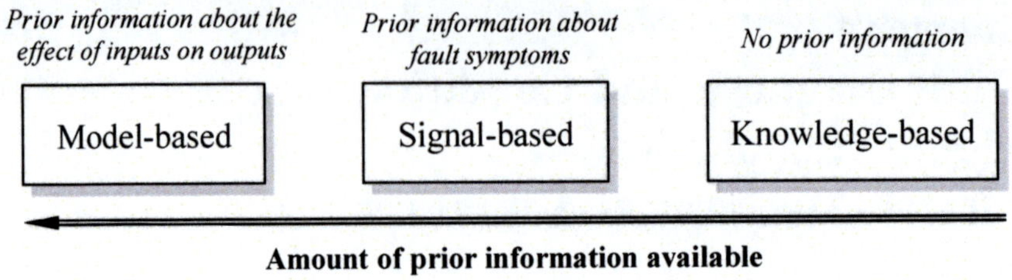

Fig. 4.1 Amount of prior information available about the faults in the different fault diagnosis approaches

- **Phase 1**
  - *Knowledge-based*: change detection method, Sect. 4.5.2;
  - *Knowledge-based*: statistical process monitoring method, Sect. 4.5.3.
- **Phase 2**
  - *Model-based*: employement of high-fidelity dynamic models for fault diagnosis, Sect. 4.3.3.

EMA for unmanned aerial vehicles. Section 4.2.4 deals with a *fault detection, isolation, prognosis* and *condition monitoring* project for primary flight control actuators for medium-altitude long-range unmanned aerial vehicles. The tested approaches include:

- *Model-based*: employement of realt-time executable dynamic models for fault diagnosis, Sect. 4.3.1.
- *Model-based*: employement of high-fidelity dynamic models for fault prognosis, Sect. 4.3.2.

Tables 4.1 and 4.2 summarize the content of the Chapter, giving a direct connection between the presented projects and solutions.

4.1 Considerations and Challenges

Fly-By-Wire (FBW) control systems and the related flight EMAs are required to be compliant with stringent safety requirements to be certified for operating in a civil environment. In order to meet these specifications, aircraft manufacturers provide high levels of hardware redundancy. This enforces the probabilities to diagnose and tolerate faults while obtaining a complete protection of the flight envelope under a wide range of external disturbances. Fault detection is mainly performed by cross checks, consistency tests, voting mechanisms, and built-in test techniques [67], see Sect. 2.3.5.

Table 4.1 Summary of the presented diagnosis solutions. (FD): Fault Detection; (FI) Fault Isolation; (CM) Condition Monitoring; (FP) Fault Prognosis

Aim	Approach	Method	Component	Fault	Section	Project
FD, FI, CM	Model based	Real-time executable dynamic models	Electrical parts	Open/short phases, motor permanent magnet degradation, decrease of voltage supply level	4.3.1	MALE UAS
FD, FP	Model based	High-fidelity dynamic models	Gear train	Jamming	4.3.2	MALE UAS
FD, FI, CM	Model based	High-fidelity dynamic models	Gear train	Jamming	4.3.3	H2020 REPRISE phase 2
FD, FI, CM	Knowledge based	Machine learning	Balls crew balls	Balls spalling	4.5.1	FP7 HOLMES
CM	Knowledge based	Change detection	Balls crew	Degradation	4.5.2	H2020 REPRISE phase 1
CM	Knowledge based	Statistical process monitoring	Ball screw	Degradation	4.5.3	H2020 REPRISE phase 1

Table 4.2 Summary of the presented diagnosis solutions grouped by objective

Aim	Approach	Method	Section
Condition monitoring (CM)	Model-based	Real-time executable dynamic models	4.3.1
		High-fidelity dynamic models	4.3.3
	Knowledge-based	Change detection	4.5.2
		Statistical process monitoring	4.5.3
Fault detection and isolation (FDI)	Model-based	Real-time executable dynamic models	4.3.1
		High-fidelity dynamic models	4.3.2 and 4.3.3
	Knowledge-based	Machine learning	4.5.1
Fault prognosis (FP)	Model-based	High-fidelity dynamic models	4.3.2

Promptly handling flight EMAs faults contributes also to the safe operation of aircraft, avoids excessive fuel consumption, and increases the aircraft autonomy. To achieve a timely diagnosis of a wider typology of faults, e.g., actuator loss of efficiency [47], more advanced methods with respect to consistency checks have to be employed. For aircraft and aerospace systems, physical redundancy will very likely be present even if analytical fault diagnosis techniques are implemented [68]. A review of the possible faults for EMAs in aerospace systems, along with the suggested analytical approach for their diagnosis, can be found in [4].

The industrial state of the art for FDI in FBW control systems is mainly based on monitoring sensor values and comparing them with thresholds. The main idea is to evaluate if the logical rule "if a signal is greater than a threshold AND if monitoring is enabled" remains true during a given time. In this case, the fault detection is confirmed and a reconfiguration is triggered. Also, the power supply is monitored. The typical configuration of the control unit architecture for a flight control consists of two dissimilar channels: the CON (control) channel and the MON (monitoring) channel. This is also the case for Airbus [23]. The flight control laws are computed separately in each channel thanks to a dedicated sensor, and any incongruence is a symptom of a fault.

Although there is still a gap between the research world and industrial practical implementation, analytical redundancy is used for the detection of a specific failure case in the A380 FBW Flight Control System (FCS) [23].

A common approach to fault diagnosis and monitoring approaches is to rely on experimental activities, where actuator response is characterized by artificially injecting faults. This allows the study of the actuator response in both normal conditions and with faults. The disadvantage of this strategy is that the time investment and rigging costs can be relevant.

The computing and memory resources for the Electronic Control Unit (ECU) of a flight EMA are limited compared to other applications. Thus, it is very difficult to use advanced processing with a high computational burden.

The typical ECU architecture with two independent CON/MON channels implies that independent clock per channel is present. A time asynchronism between the two channels can grow if some data must be exchanged from one channel to the other. Furthermore, not all functions are computed with the same clock period even inside the same ECU.

In order to obtain certification, it is necessary to demonstrate that the probability of all catastrophic failures is very low (less than 10^{-9} per flight hour). It follows that the probability of not detecting these faults should be less than 10^{-9} per flight hour.

4.2 Relevant Recent Aerospace Projects

This section presents four real-world projects that were carried out in the context of European/national research and industrial activities. The aim is to present recent experimental projects and architectural solution for fault diagnosis and condition monitoring approaches for electro-mechanical actuators in the aerospace environment. The methodologies employed in solving these diagnosis problems are then presented in the next Sects. 4.3–4.6.

4.2.1 FP7 HOLMES Project

The Health On Line Monitoring for Electromechanical actuator Safety (HOLMES) project [37–39] was funded by the CleanSky Joint Technology Initiative (JTI), within the 7-th Framework Program (FP7) of European Union (EU), see Fig. 4.2, and saw a collaboration between University of Bergamo, Liebherr Aerospace and Umbra-Group.

Objectives. The aim of the HOLMES project is to test and devise a fault detection approach for secondary surfaces EMA of large airliners. The main focus of the detection is related to recirculating balls of ball screw transmissions.

The RTCA DO-160 "Environmental Conditions and Test Procedures for Airborne Equipment" standard RTCA DO-160 [55] has been consulted in order to evaluate the susceptibility of the actuator to external conditions (e.g. temperature and altitude variations, humidity, operational shock, vibration, explosive atmosphere, waterproofness, etc.)

By considering the conditions that were feasible to test and those that had little effect on the actuator measured signals, we chose to perform, in addition to environmental temperature tests, also low temperature tests.

Diagnosis Methods. In the context of the HOLMES project, a fault detection and isolation method, based on supervised machine learning algorithms, is presented in Sect. 4.5.1.

European Commission | Clean Sky

Fig. 4.2 The HOLMES project was funded under the CleanSky Joint Technology Initiative, FP7 European research program FP7-JTI-CS, Grant Agreement number 338549

4.2.1.1 Identification of the Most Critical Failures

The considered actuator is composed of a five-phase Brushless BLDC machine with a direct-drive ball screw transmission, Fig. 4.3. The focus of the project is on the transmission component. Figure 4.4 summarizes the most critical faults for a geared and direct-drive EMA in the context of the project. The next results will focus only on the direct-drive specimen. In this case, the most critical faults highlighted by the analysis are:

- recirculation jam;
- spalling on raceway;
- broken balls inside the transmission nut. This will be the focus on the knowledge-based method employed in Sect. 4.5.1.

4.2.1.2 Experimental Setup

EMA and test bench design. The test bench, depicted in Fig. 4.5, is able to test up to two parallel EMAs. Only one of them is used for the tests. The employed five-phases BLDC machine is intrinsically fault tolerant, see Sect. 2.2. The direct-drive ball screw transmission is equipped with a nut for converting the rotational motion to a linear one, see Sect. 1.2.2.3. The nut under test consists of two channels for the recirculation of the steel balls.

Fig. 4.3 The direct-drive five-phases EMA under consideration in the HOLMES project

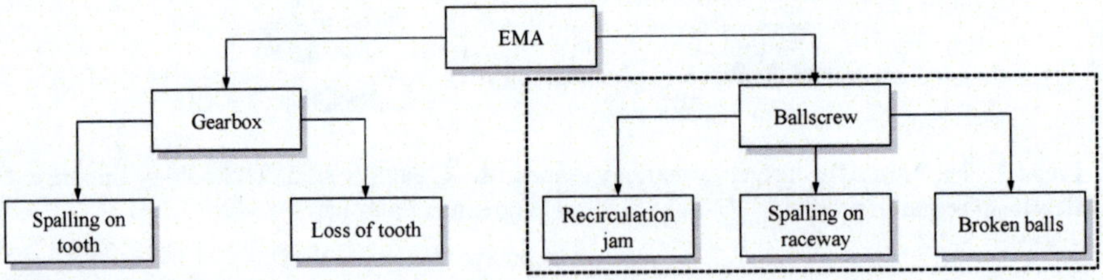

Fig. 4.4 Failure modes studied in the HOLMES project. Next sections will focus on the broken balls fault

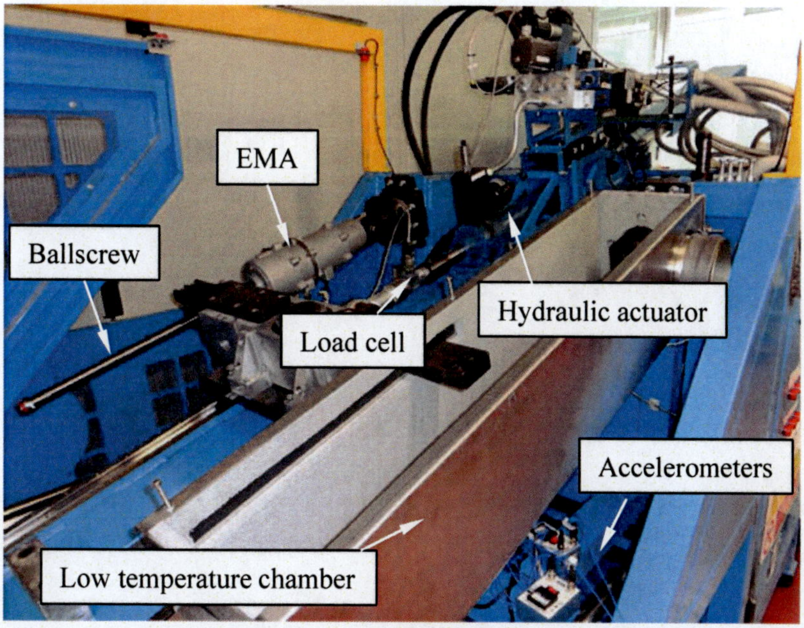

Fig. 4.5 HOLMES project test rig with main components. Chamber for low temperature tests is visible

The actuator is controlled by a cascade closed-loop control system, involving position, speed, and current control loops. The EMA stroke ranges from 0 mm to about 400 mm. The load force acting on the EMA is generated by a controlled hydraulic cylinder [9].

Collected and available measurements.
The measured variables (Fig. 4.6) consist of:

- EMA position reference, actual position, and speed (sampling frequency 5 kHz);
- Phase A, B, C, D, E currents (sampling frequency 5 kHz);
- hydraulic load reference and load cell measure (sampling frequency 1 kHz);
- hydraulic cylider position and speed (sampling frequency 1 kHz);
- hydraulic servovalve reference and actual value (sampling frequency 1 kHz);
- cage temperature of the EMA (measured by a K-type thermocouple);
- nut vibrations, measured by n.2 accelerometers orthogonally placed on the nut (sampling frequency 20 kHz).

Fault implementation and test conditions. An Electrical Discharge Machine (EDM) was used to inject the spalling of steel balls fault. This operation removes material from the balls, providing a truncated sphere shape, see Figs. 4.7 and 4.8.

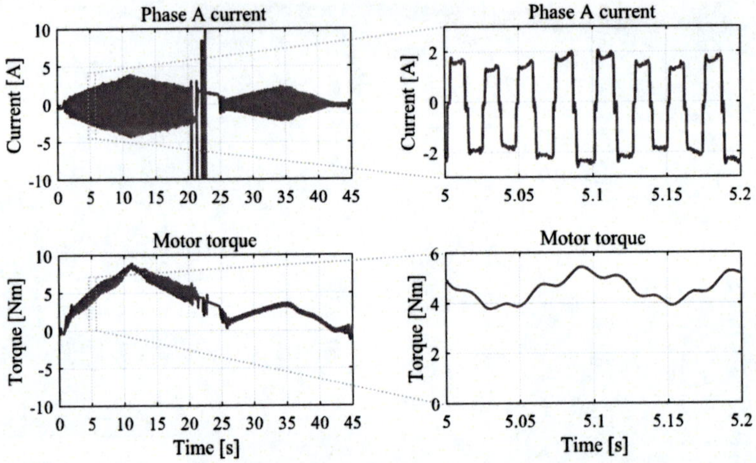

Fig. 4.6 Current of phase A and detail (top). Estimated torque and detail (bottom). Reproduced from [38] by permission. © IFAC 2017

(a) Light damaged vs. healthy balls

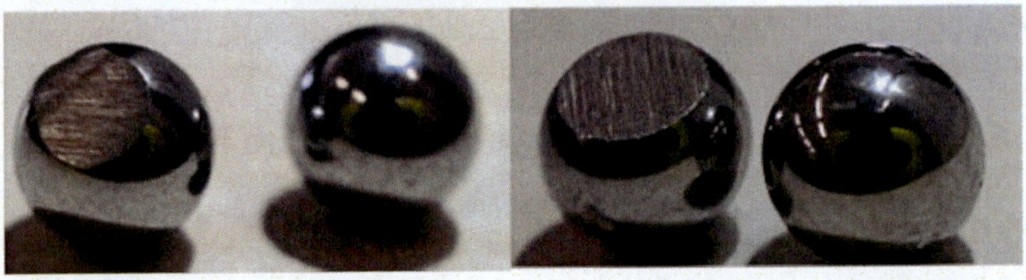

(b) Medium damaged vs. healthy balls. (c) Heavy damaged vs. healthy balls.

Fig. 4.7 Qualitative comparison of the faults injected on the ball screw balls

Fig. 4.8 Quantification of the injected faults on ball screw balls. Reproduced from [38] by permission. © IFAC 2017

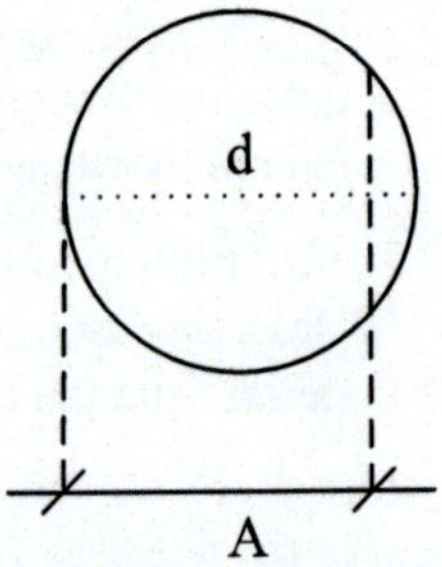

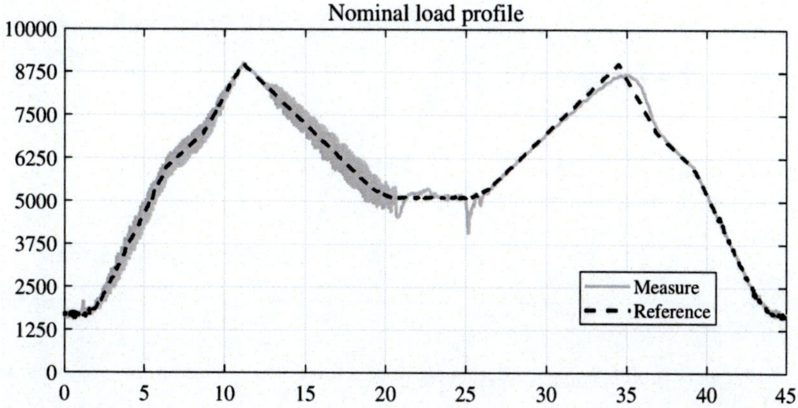

Fig. 4.9 Nominal load profiles employed during the test sessions. Reproduced from [38] by permission. © IFAC 2017

The nominal balls diameter is $d = 3.5$ mm. For faulty balls, we have that

- *Light damage*: $A = 3.3$ mm;
- *Medium damage*: $A = 3.2$ mm;
- *High damage*: $A = 3.1$ mm.

Thus, the following four conditions were tested:

1. *Healty condition*: no damaged balls;
2. *Light faulty state*: 6 light damaged, 6 medium damaged and 6 high damaged balls for each recirculation channel;
3. *Medium faulty state*: 20 high damaged balls for each recirculation channel;
4. *Severe faulty state*: 40 high damaged balls for each recirculation channel.

The external EMA temperature was kept constant between one test and another, to minimize the effect of the temperature on the EMA behavior. Furthermore, low temperature tests (that reached −40 °C) were performed by means of a cold chamber built around the EMA (Fig. 4.5).

Test profiles. The chosen load profiles are shown in Fig. 4.9, where non-idealities in tracking are due to test bench limitations. They consist of:

- a typical position-dependent high lift load profile;
- a constant load of 12 kN;
- a constant load of 15 kN.

The EMA motion profile has been defined as follows, see Fig. 4.10:

- position run from 0 mm to 411 mm in 20 s;
- velocity ramp from $0\frac{\text{mm}}{\text{s}}$ to $21\frac{\text{mm}}{\text{s}}$ in 2 s.

For each test, we performed two runs of the motion profile.

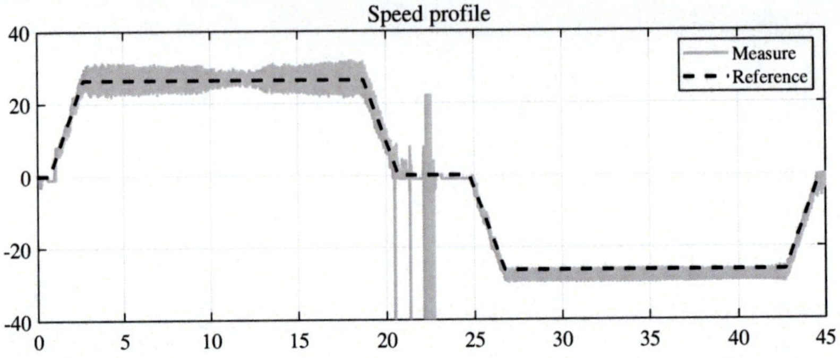

Fig. 4.10 Speed profile employed during the test sessions. Reproduced from [38] by permission. © IFAC 2017

4.2.2 H2020 REPRISE Project: Phase 1

This section presents the Reliable Electro-mechanical actuator for PRImary SurfacE with health monitoring (REPRISE) project.[2] Phase 1 saw a collaboration between the University of Bergamo, UmbraGroup, Piaggio Aerospace, and Zettlex ltd. See Sect. 4.2.3 for the second phase of the project.

The REPRISE project aims to improve the Technological Readiness Level (TRL) of flight control systems based on EMAs for small aircrafts. In particular, the EMA is devoted to the control of the primary flight surfaces (ailerons, rudder, and elevators), Fig. 4.11. The project is structured in two parts:

- developing a Condition Monitoring (CM) system on an already existent EMA architecture (*Phase 1*);
- designing a new electro-mechanical actuator architecture and test (or redesign) the final monitoring solution (*Phase 2*).

In the following, we present the results of Phase 1 of the project.
Phase 1 Objectives. The first phase of the project is devoted to [40, 43]:

- test the actuator with an experimental endurance campaign using a test bench;
- develop a condition monitoring system to detect deteriorations of the ball screw components.

Diagnosis Methods. In the context of phase 1 of the REPRISE Project, a *condition monitoring* approach based on a change-point detection algorithm is presented in Sect. 4.5.2, while a *monitoring strategy* based on statistical process monitoring is presented in Sect. 4.5.3.

---

[2]This project has received funding from the Clean Sky 2 Joint Undertaking under the European Union's Horizon 2020 research and innovation programme under grant agreement 717112, call: JTI-CS2-2015-CFP02-SYS-03-01.

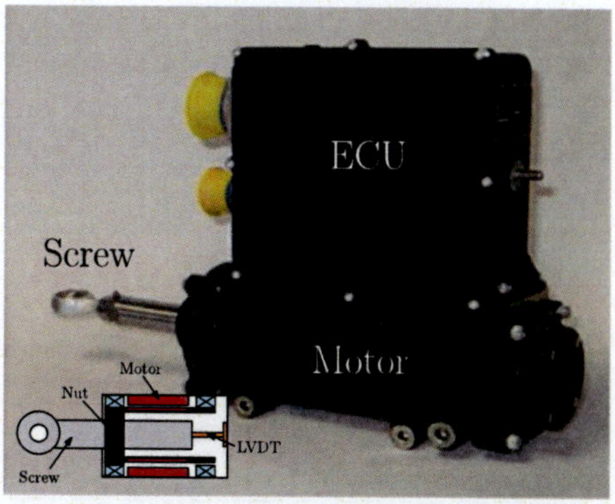

Fig. 4.11 REPRISE EMA general view. Reproduced from [40]—originally published open access and licensed under CC-BY 4.0. https://ieeexplore.ieee.org/stamp/stamp.jsp?tp=&arnumber=8878102

Table 4.3 Position configurations in the experimental table. Reproduced from [40]—originally published open access and licensed under CC-BY 4.0. https://ieeexplore.ieee.org/stamp/stamp.jsp?tp=&arnumber=8878102

Mission phase	Percentage of mission duration	Temperature (C°)
A/C on ground and on takeoff	5	+70
A/C on ground and on takeoff	2	+85
Climb	6	−15
Transfer and loiter	78	−45
Descent	6	−15
Landing	3	+50

4.2.2.1 Critical Failures Selection

As a first step, a Failure Mode, Effects and Criticality Analysis (FMECA) was performed to select the most critical EMA failures. Following the FMECA (see Sect. 2.3.4), a Fault Tree Analysis (FTA) (see Sect. 2.3.3) is conducted to check the actuator compliance with the requirements about safety.

Failure Mode Effect and Criticality Analysis. The FMECA identified 1950 failure modes. The MIL-HDBK-217F handbook [1] was used to evaluate the failure mode rates of each single component, over the one million flight hours mission profile reported in Table 4.3.

Table 4.4 shows the resulting Failure Mode Effect Summary (FMES), see Sect. 2.3.1. FMES results denote how, in most cases, component failures lead to a "no functional effect" outcome. Thus, they do not lead to critical failures.

Fault Tree Analysis. The results in Table 4.4 are given in input to the FTA, and results are reported in Table 4.5, by considering 1 Flight Hour (FH) of operational time. The FTA (see Sect. 2.3.3) was performed to evaluate the compliance of the EMA with respect to predefined safety requirements. The FTA focused on four

Table 4.4 Failure mode effect summary for REPRISE phase 1 EMA. Reproduced from [40]—originally published open access and licensed under CC-BY 4.0. https://ieeexplore.ieee.org/stamp/stamp.jsp?tp=&arnumber=8878102

End effect	Failures per million hours (fpmh)
Actuator jam	3.647×10^{-2}
Actuator runaway	6.000×10^{-6}
False alarm signal	7.859×10^{-2}
Loss of actuator	6.152
Loss of capability to engage the static brake	3.664×10^{-1}
Loss of service communication	5.748×10^{-2}
No functional effect	1.039×10^{1}
No functional effect. The failure could become critical in presence of other failures	1.777×10^{-1}
No significant effect	4.366×10^{-2}
Possible loss of actuator	1.550×10^{-4}
Static brake always engaged	6.660×10^{-2}

Table 4.5 Fault tree analysis summary for REPRISE phase 1 EMA. Reproduced from [40]—originally published open access and licensed under CC-BY 4.0. https://ieeexplore.ieee.org/stamp/stamp.jsp?tp=&arnumber=8878102

Top FTA event	Risk likelihood (1/FH)	Requirement (1/FH)
Actuator loss of control/function	6.218×10^{-6}	$< 10^{-7}$
Actuator free floating	6.000×10^{-9}	$< 10^{-7}$
Actuator runaway	2.0709×10^{-12}	$< 10^{-8}$
Actuator jam	3.648×10^{-8}	$< 10^{-9}$

fundamental failure modes, with the associated risk likelihood (see also Sect. 3.3 for the terminology):

1. Actuator loss of control/function;
2. Actuator free floating;
3. Actuator runaway;
4. Actuator jam.

The FTA summary results are reported in Table 4.5. One can notice that the actuator loss of control/function and jam failure requirements are not satisfied. These events depend mainly on damages to the the transmission, i.e., the ball screw. One of the most plausible causes of the ball screw damaging is the lack of adequate lubricant: this notion will be taken into consideration for the design of the experimental conditions.

Performance requirements and design of experiments. The considered EMA has been designed according to the following specifications:

Table 4.6 Aileron duty cycle for REPRISE phase 1 EMA. Reproduced from [40]—originally published open access and licensed under CC-BY 4.0. https://ieeexplore.ieee.org/stamp/stamp.jsp?tp=&arnumber=8878102

Amplitude (% full range: 20.5 *mm*)	Load (% max load: 1346 N)	Cycles (per mission)	Cycles (total)
0.5	2.5	250.0	2.17×10^5
1	1	4370.5	3.80×10^6
1	5	100.0	8.70×10^4
2.5	2.5	1984.0	1.73×10^6
5	5	40.0	3.48×10^4
5	25	2.0	1.74×10^3
10	10	343.5	2.99×10^5
25	10	0.0	4.35×10^0
25	25	12.7	1.10×10^4
50	25	4.0	3.48×10^3
50	50	0.2	1.74×10^2

- *control performance*;
- *physical dimensions and total mass*;
- *nominal operating conditions*, i.e., compliance respect to predefined *duty cycles* for each of the actuated primary flight control surface (aileron, elevator, rudder).

These duty cycles are reported in Tables 4.6, 4.7 and 4.8, where:

- *Full range*: surface full deflection (Aileron: 20.5 m*m*, Elevator: 48.5 m*m*, Rudder: 47.4 m*m*);
- *Max load*: largest rated *axial load* (Aileron: 1346 N, Elevator: 1405 N, Rudder: 1494 N). The EMA has to tolerate a *radial load* up to 17% of the axial load;
- the EMA is required to perform such duty cycles, with a *sinusoidal position profile* with amplitude given by the *amplitude* column of the Tables 4.6, 4.7 and 4.8, and a frequency between 0.2 and 1 H*z*.

It can be seen that the rudder configuration is the most demanding condition, considering the axial force and strokes. Thus, the rudder configuration is taken as a guide to design the experimental campaign.

These specification tables also show that the EMA is required to almost always actuate a small stroke (less than 1 m*m*) with a very small load (less than 150 N). Experimental tests will, however, use not only these small strokes, but also higher sinusoidal amplitudes in order to stress the system and accelerate its degradation.

Table 4.7 Elevator duty cycle for REPRISE phase 1 EMA. Reproduced from [40]—originally published open access and licensed under CC-BY 4.0. https://ieeexplore.ieee.org/stamp/stamp.jsp?tp=&arnumber=8878102

Amplitude (% full range: 20.5 mm)	Load (% max load: 1346 N)	Cycles (per mission)	Cycles (total)
0.5	2.5	250.0	2.17×10^5
1	2.5	4370.5	3.80×10^6
1	5	100.0	8.70×10^4
2.5	2.5	100.0	8.70×10^4
2.5	5	1884.0	1.64×10^6
5	5	40.0	3.48×10^4
5	25	1.0	8.70×10^2
10	10	0.4	3.48×10^2
10	25	171.2	1.49×10^5
25	50	7.4	6.39×10^3
50	25	4.0	3.48×10^3
50	50	1.0	8.70×10^2

Table 4.8 Rudder duty cycle for REPRISE phase 1 EMA. Reproduced from [40]—originally published open access and licensed under CC-BY 4.0. https://ieeexplore.ieee.org/stamp/stamp.jsp?tp=&arnumber=8878102

Amplitude (% full range: 20.5 mm)	Load (% max load: 1346 N)	Cycles (per mission)	Cycles (total)
0.5	10	250.0	2.17×10^5
1	5	4370.5	3.80×10^6
2.5	5	100	8.70×10^4
2.5	50	2.0	1.74×10^3
5	10	0.2	1.74×10^2
5	25	343.3	2.98×10^5
10	50	12.7	1.10×10^4
50	25	4.0	3.48×10^3
50	100	0.2	1.78×10^2

4.2.2.2 Experimental Setup

EMA and test bench design. The EMA is equipped with a 3-phases 5 poles BLDC motor, supplied by two 28 Vdc power lines. Three Hall sensors and an embedded LVDT transducer provide the motor position. The EMA has a stroke of $\pm 30\,mm$ from its homing position (position offset equal to 0 mm). The ball screw transmission presents 8 circuits with 1 turn each. An anti-rotation device provides the EMA with the ability to compensate for small radial loads.

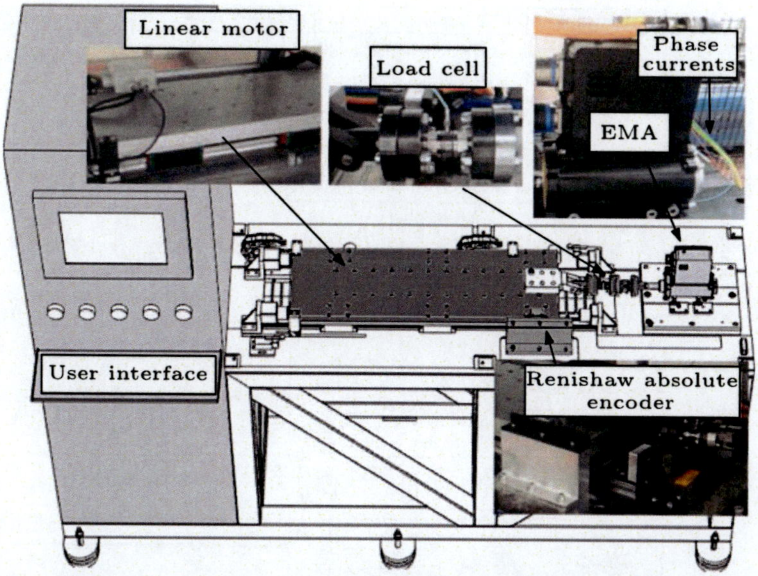

Fig. 4.12 REPRISE project test bench with main components. Reproduced from [40]—originally published open access and licensed under CC-BY 4.0. https://ieeexplore.ieee.org/stamp/stamp.jsp?tp=&arnumber=8878102

The EMA is positioned within a bench, inside an air-cooled box, where a linear motor provides the test loads, see Fig. 4.12. The bench permits also to monitor the system status with additional sensors, such as an absolute optical encoder to measure the EMA absolute position and phase current sensors to measure the 3 motor phase currents.

Collected and available measurements. The measured physical variables are:

1. load reference of the linear motor;
2. EMA box temperature;
3. current supplied to the EMA (from power supply);
4. load cell measure;
5. position reference of the EMA;
6. position measure of the EMA from (embedded) LVDT sensor;
7. position measure of the EMA from (external) absolute optical sensor;
8. current supplied to the linear motor (from its drive);
9. n.3 phase currents of the EMA.

The phase currents are sampled at 4800 Hz, while other variables are acquired at 100 Hz. Figure 4.13 shows the behavior of some measurements, with position reference sine frequency of 0.5 Hz, amplitude of 10 mm, and load of 300 N.

Testing procedure. The goal of the tests are:

1. induce degradations on the ball screw;

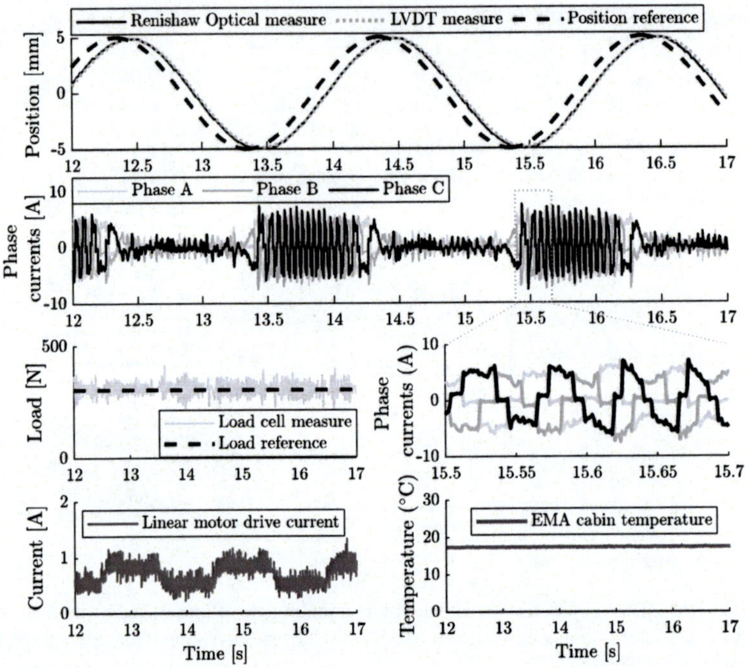

Fig. 4.13 Examples of measurements from the test bench. Reproduced from [40]—originally published open access and licensed under CC-BY 4.0. https://ieeexplore.ieee.org/stamp/stamp.jsp?tp=&arnumber=8878102

2. provide measurements for monitoring the health state of the ball screw.

In order to obtain these goals, two types of test trials were used:

- *Monitoring trials*, in order to characterize the EMA health status;
- *Endurance trials*, to deteriorate the ball screw components by endurance tests, with load values that exceeded the design allowable ones.

The experimental campaign alternated Monitoring and Endurance trials.

The EMA under test can be commanded only by sinusoidal position references. So, an experiment is completely defined by the frequency, amplitude, and offset of the position reference trajectory. A total of 10 frequency values are used: {0.1, 0.3, 0.5, 0.8, 0.9, 1, 1.5, 2, 2.5, 4} (in Hz). Table 4.9 describes the position configurations tested for each of these frequencies. So, an *experimental session* consists of repeating the 6 test configurations of Table 4.9 for each of the 10 frequency values defined above. Most of the time, a Monitoring trial consists of executing one experimental session, while Endurance trials repeat many successive identical experimental sessions.

Monitoring trials were performed with 300 N load, which resulted in a condition where no ball is over-stressed (*nominal H0 condition*). Endurance trials were performed with a chosen load of 800 N (*over-stressed H1 condition*), be means of a Finite Element Method (FEM) analysis on the ball screw.

Test conditions. The following actions have been undertaken to accelerate the ball screw/nut assembly degradation process:

1. only 3 circuits out of 8 in the ball screw were employed: in this way, the remaining circuits will carry higher loads;
2. the EMA anti-rotation device was removed, to stress more the balls and the ball screw tracks;
3. a radial load equal to 17% of the axial load was applied;
4. the lubricant inside the ball screw/nut assembly was progressively removed.

The lack of lubricant is one of the main causes of ball screw degradation. Thus, the following operating conditions were tested:

1. *Standard level of lubricant*;
2. *Poor level of lubricant*: about half removed;
3. *Lubricant completely removed*.

Test report. Table 4.10 reports the number of screw revolutions at different loads and lubricant conditions, considering the tests performed after the anti-rotation removal. After an initial phase where no degradation effects were observed, this device was removed. Figure 4.14 presents a complete test summary where particular Monitoring trial dates are highlighted.

4.2.3 H2020 REPRISE Project: Phase 2

This section presents the aims and activities regarding the Phase 2 of the REPRISE project (see Sect. 4.2.2 for a description of the first phase).
Phase 2 objectives. The REPRISE Project—Phase 2 was carried out by Umbra-Group, in collaboration with the University of Pisa and AESIS srl, by pursuing three main objectives:

Table 4.9 Position configurations in the experimental table. Reproduced from [40]—originally published open access and licensed under CC-BY 4.0. https://ieeexplore.ieee.org/stamp/stamp.jsp?tp=&arnumber=8878102

Offset position	Amplitude position	Stroke range	N° of cycles
0	5	[−5, +5]	100
0	10	[−10, +10]	100
+10	5	[+5, +15]	100
+10	10	[0, +20]	100
−10	5	[−15, −5]	100
−10	10	[−20, 0]	100

- the architecture definition, the design and the manufacturing of a fault-tolerant / jamming-tolerant EMA for safety-critical aerospace applications;
- the design and the verification at TRL 4 of fault-tolerant control laws, capable of reconfiguring the EMA operation mode in case of major faults while maintaining adequate dynamic performances;
- the design and the verification at TRL 3 of condition monitoring algorithms, with particular reference those dedicated to jamming faults.

Diagnosis Methods. In the second phase of the REPRISE project, the fault diagnosis and condition monitoring is applied by means of a model-based approach, aiming to assess jamming faults in the ball screw transmission, see Sect. 4.3.3.

Table 4.10 Number of screw revolutions after anti-rotation removal. Reproduced from [40]—originally published open access and licensed under CC-BY 4.0. https://ieeexplore.ieee.org/stamp/stamp.jsp?tp=&arnumber=8878102

	Standard level of lubricant	Poor level of lubricant	Removed lubricant	Total
Condition H0: 300 N	185.609	250.388	333.023	769.020
Condition H1: 800 N	146.846	145.212	371.579	663.637
Total	332.455	395.600	704.602	1.432.657

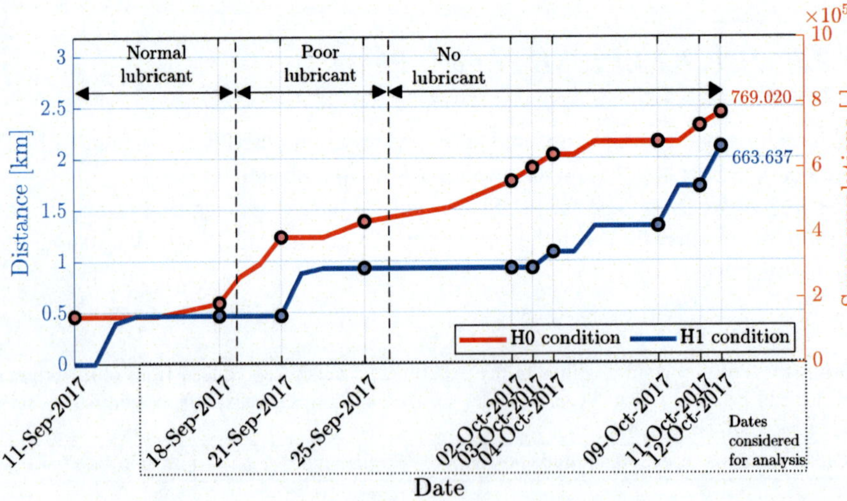

Fig. 4.14 REPRISE Phase 1: performed tests summary. Reproduced from [40]—originally published open access and licensed under CC-BY 4.0. https://ieeexplore.ieee.org/stamp/stamp.jsp?tp=&arnumber=8878102

4.2.3.1 Electro-Mechanical Actuator Description

The actuator developed by UmbraGroup for the REPRISE project – Phase 2 is a fault-tolerant/jamming-tolerant EMA (Fig. 4.15a), composed of:

- dual-redundant three-phase Brushless AC machines (BLACM)s, driven via Field Oriented Control (FOC) technique and by a Space-Vector Pulse-Width Modulation (SVPWM) method;
- dual ECU with CON-MON architecture, implementing the condition monitoring algorithms and the closed-loop control functions (based on nested loops on motors' currents, motors' speed, and output position);
- patented jamming-tolerant mechanical transmission with differential ball screws (Fig. 4.15b), in which the two motors engage, via integrated ball-nuts, an intermediate screwshaft having three threaded portions: two external ones for the motors' ball-nuts, and an internal one for the output shaft.

The structure of the control and monitoring boards of the fault-tolerant EMA is schematically depicted in Fig. 4.16. The EMA ECU is essentially composed of:

- n. 2 control modules (CON_1, CON_2), implementing the closed-loop control algorithms on motors' current, motors' shaft speed, and output shaft position;
- n. 2 monitor modules (MON_1, MON_2), implementing the health monitoring algorithms;
- n. 12 current sensors, two per each phase of the two motors;
- n. 4 resolvers, two per motor, to sense the motors' rotations;
- n. 2 cone-type proximity sensors, to sense the screw shaft translation;
- n. 2 Linear Variable Differential Transformer (LVDT) sensors, to sense the output shaft position.

The MON_i module ($i = 1, 2$) interacts with:

- the Flight Control Computer (FCC) through a RS422 standard serial communication;
- the CON_i module through a Serial Peripheral Interface(SPI) bus protocol;

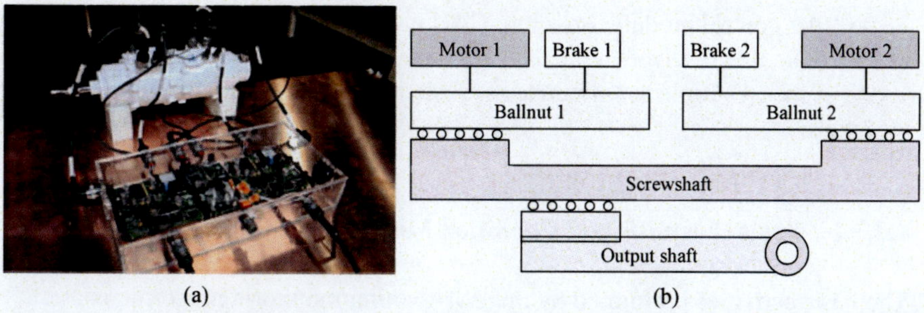

Fig. 4.15 Fault-tolerant EMA with differential ball screws: **a** prototype rigging; **b** kinematic concept (patented by UmbraGroup)

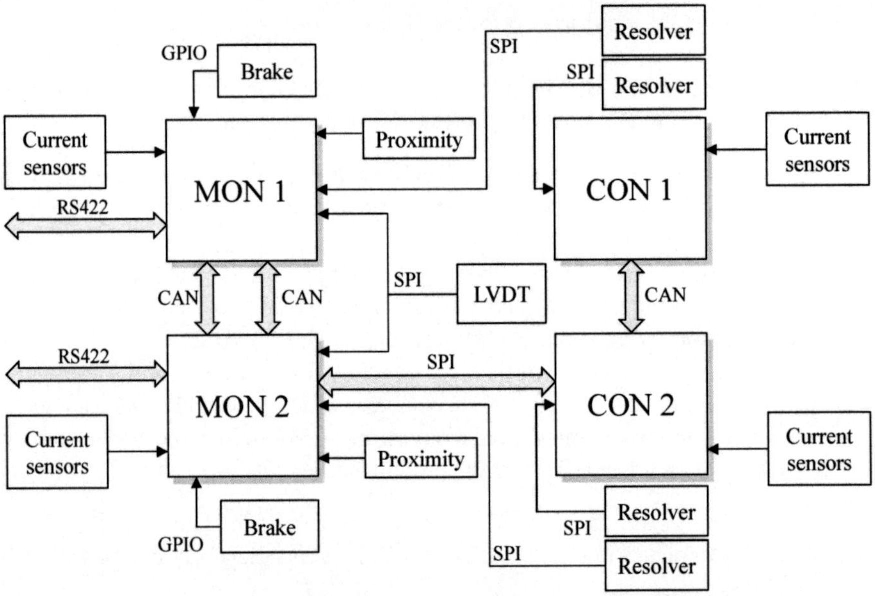

Fig. 4.16 Control and monitoring electronic board diagram

- the other monitor module through a redundant Controller Area Network (CAN) bus protocol;
- one cone-type proximity sensor through an analogic signal;
- one of the two resolvers related to i-th motor through an SPI bus protocol;
- one of the two LVDT sensors through an SPI bus protocol;
- one of the two triple set of current sensors related to i-th motor through analogic signals;
- the brake related to i-th motor through a General-Purpose Input/Output (GPIO) interface.

The CON_i module ($i = 1, 2$) interacts with:

- the MON_i module through an SPI bus protocol;
- the other control module through a CAN bus protocol;
- one of the two resolvers related to i-th motor through an SPI bus protocol;
- one of the two triple set of current sensors related to i-th motor through analogic signals.

4.2.3.2 Fault Diagnosis and Condition Monitoring System

The basic activities performed by the EMA voting/monitoring algorithms are:

- *Data voting*: i.e., the process of obtaining a unique consolidated value of a data from multiple measurements or estimations of it;

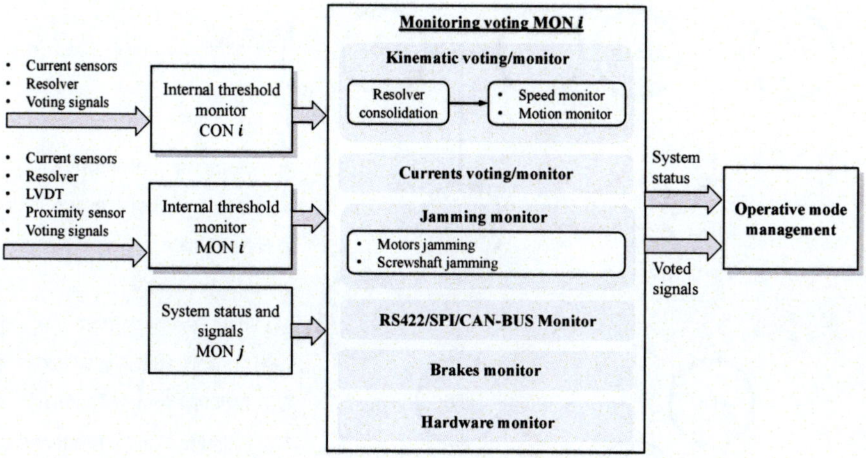

Fig. 4.17 MON_i module block diagram ($i = 1, 2$)

- *Fault detection*: i.e., the process of identifying a system malfunction or a deviation from expected system behavior, carried out by processing system measurements or estimations;
- *Fault isolation*: i.e., the process of determining the fault mode that is responsible for the deviation from expected system behavior;
- *Condition monitoring*, the process of evaluating the system health-state by incrementing/decrementing of specific numerical counters.
- *Fault compensation*: i.e., the process of actively intervening to modify the system configuration after a fault, aiming to recover some level of system performance;
- *Fault accomodation*: i.e., the adaptation of the control laws when a major fault is detected.

The diagnosis architecture developed for each MON_i module ($i = 1, 2$) is schematically reported in Fig. 4.17, where it is worth noting that the following monitoring algorithms are executed:

- *kinematic voting/monitor* (Sect. 4.2.3.3), composed of three sub-functions:

  a. speed monitor, for the detection of motor overspeed due to hardover faults;
  b. resolvers' consolidation;
  c. motion monitor;

- *currents voting/monitor* (Sect. 4.2.3.4);
- *jamming monitor* (Sect. 4.3.2);
- *RS422/SPI/CAN-bus monitor*;
- *brakes monitor*;
- *hardware monitor*.

The motion and current monitors are briefly described in next sections. Then, more details will be given on the jamming monitoring algorithm in Sect. 4.3.2.

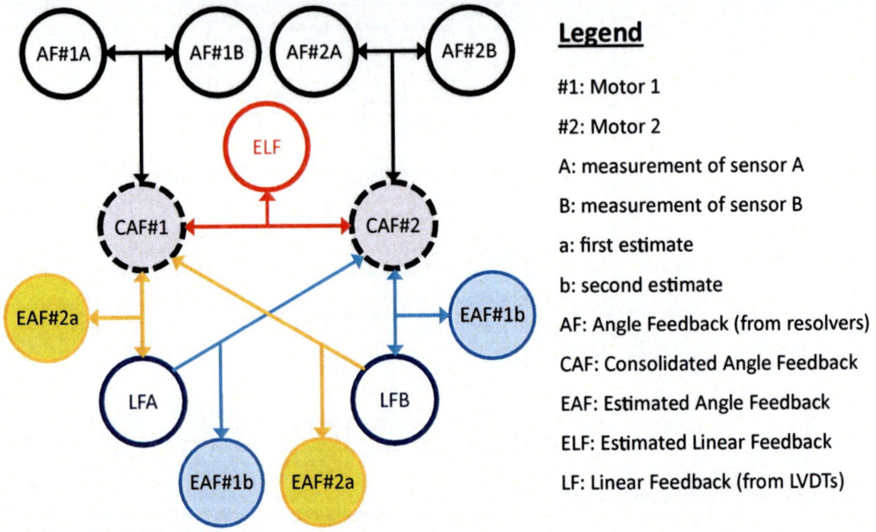

Fig. 4.18 Motion monitor: logic of the analytical redundancy on positions' sensing

4.2.3.3 Motion Monitor

The Motion Monitor computes voted values of motors' rotation, screw shaft position and output shaft position by using triple redundant signals, obtained by combining the dual sensors' measurements with *analytical reconstructions* derived from kinematic relationships. The mechanical transmission is based on a speed-summing arrangement and the output displacement $x_o(t)$ is related to motors' rotations $\theta_1(t)$ and $\theta_2(t)$ via

$$x_o(t) = b_1\theta_1(t) + b_2\theta_2(t), \tag{4.1}$$

where b_1 and b_2 depend on the pitches of the three nut-screw couplings. By using (4.1), once that two out of the three quantities $x_o(t)$, $\theta_1(t)$ and $\theta_2(t)$ are known, the remaining one can be estimated.

Taking into account that the EMA integrates two LVDTs and that each motor integrates two resolvers providing a consolidated rotation signal (Resolver Consolidation function in Fig. 4.17), five position estimates can be calculated from Consolidated Angle Feedback (CAF) and Linear Feedback (LF) measurements, as described in (4.2):

$$\text{EAF\#1a} = (\text{LFA} - b_2\text{CAF\#2})/b_1 \tag{4.2a}$$

$$\text{EAF\#1b} = (\text{LFB} - b_2\text{CAF\#2})/b_1 \tag{4.2b}$$

$$\text{EAF\#2a} = (\text{LFA} - b_1\text{CAF\#1})/b_2 \tag{4.2c}$$

$$\text{EAF\#2b} = (\text{LFB} - b_1\text{CAF\#1})/b_2 \tag{4.2d}$$

$$\text{ELF} = b_1\text{CAF\#1} + b_2\text{CAF\#2}, \tag{4.2e}$$

Table 4.11 Motion Monitor: EMA states resulting from position sensors' faults. CAF: Consolidated Angle Feedback; LFA: Linear Feedback sensor A; LFB: Linear Feedback sensor B; VAF: Voted Angle Feedback; VLF: Voted Linear Feedback

Number of faults	CAF#1	CAF#2	LFA	LFB	VAF#1	VAF#2	VLF	EMA state
0	OK	OK	OK	OK	CAF#1	CAF#2	LFA	Operative
					EAF#1a	EAF#2a	LFB	
					EAF#1b	EAF#2b	ELF	
1	Fail	OK	OK	OK	EAF#1a	CAF#2	LFA	Operative
					EAF#1b		LFB	
	OK	Fail	OK	OK	CAF#1	EAF#2a	LFA	Operative
						EAF#2b	LFB	
	OK	OK	Fail	OK	CAF#1	CAF#2	LFB	Operative
					EAF#1b	EAF#2b	ELF	
	OK	OK	OK	Fail	CAF#1	CAF#2	LFA	Operative
					EAF#1a	EAF#2a	ELF	
2	Fail	OK	Fail	OK	EAF#1b	CAF#2	LFB	Operative
	Fail	OK	OK	Fail	EAF#1a	CAF#2	LFA	Operative
	OK	Fail	Fail	OK	CAF#1	EAF#2b	LFB	Operative
	OK	Fail	OK	Fail	CAF#1	EAF#2a	LFA	Operative
	OK	OK	Fail	Fail	CAF#1	CAF#2	ELF	Operative
	Fail	Fail	OK	OK	*No signal*	*No signal*	LFA	Fail-safe
							LFB	

Figure 4.18 explains the acronyms and the rationale of the position estimates.

Thanks to the analytical redundancy strategy, the EMA operates with *triple redundancy on each required position signal*, so that the system is capable of tolerating up to two faults to any position sensor, except the fault of two resolvers on different motors, see Table 4.11.

4.2.3.4 Currents Voting/Monitor

This monitor algorithm computes voted values of currents for each phase of the two BLACMs by using the dual redundant signals derived from currents sensors.

In addition to the voting function, the monitor detects and isolates the current sensors faults and coil faults (open circuit, short circuits, etc.). In the normal operation of a three-phase BLACM, the currents' sum is constant (near to 0). Thus, by evaluating this signal, it is possible to check the correct behavior of the coil and/or the correct measurement of the related sensor.

Table 4.12 Groups and classes definition for the currents' monitor

Group	Currents sums	Common measurements	Class
G_{Aa}	$\boldsymbol{\Sigma}_{if1}, \boldsymbol{\Sigma}_{if2}, \boldsymbol{\Sigma}_{if3}, \boldsymbol{\Sigma}_{if4}$	$i_{af\mathrm{A}}$	C_{a}
G_{Ba}	$\boldsymbol{\Sigma}_{if5}, \boldsymbol{\Sigma}_{if6}, \boldsymbol{\Sigma}_{if7}, \boldsymbol{\Sigma}_{if8}$	$i_{af\mathrm{B}}$	
G_{Ab}	$\boldsymbol{\Sigma}_{if1}, \boldsymbol{\Sigma}_{if2}, \boldsymbol{\Sigma}_{if5}, \boldsymbol{\Sigma}_{if6}$	$i_{bf\mathrm{A}}$	C_{b}
G_{Bb}	$\boldsymbol{\Sigma}_{if3}, \boldsymbol{\Sigma}_{if4}, \boldsymbol{\Sigma}_{if7}, \boldsymbol{\Sigma}_{if8}$	$i_{bf\mathrm{B}}$	
G_{Ac}	$\boldsymbol{\Sigma}_{if1}, \boldsymbol{\Sigma}_{if4}, \boldsymbol{\Sigma}_{if6}, \boldsymbol{\Sigma}_{if7}$	$i_{cf\mathrm{A}}$	C_{c}
G_{Bc}	$\boldsymbol{\Sigma}_{if2}, \boldsymbol{\Sigma}_{if3}, \boldsymbol{\Sigma}_{if5}, \boldsymbol{\Sigma}_{if8}$	$i_{cf\mathrm{B}}$	

Starting from the six available current measurements, eight combinations of currents' sums $\boldsymbol{\Sigma}_{if} \in \mathbb{R}^{8\times 1}$ can be obtained as in (4.3):

$$\boldsymbol{\Sigma}_{if} = \begin{bmatrix} 1\,1\,1\,0\,0\,0 \\ 1\,1\,1\,0\,0\,0 \\ 1\,0\,0\,0\,1\,1 \\ 1\,0\,1\,0\,1\,0 \\ 0\,1\,0\,1\,0\,1 \\ 0\,1\,1\,1\,0\,0 \\ 0\,0\,1\,1\,1\,0 \\ 0\,0\,0\,1\,1\,1 \end{bmatrix} \cdot \begin{bmatrix} i_{af\mathrm{A}} \\ i_{bf\mathrm{A}} \\ i_{cf\mathrm{A}} \\ i_{af\mathrm{B}} \\ i_{bf\mathrm{B}} \\ i_{cf\mathrm{B}} \end{bmatrix} \tag{4.3}$$

As reported in Table 4.12, by considering the contribution of each current measurement to each single element $\boldsymbol{\Sigma}_{ifh} \in \mathbb{R}$ of $\boldsymbol{\Sigma}_{if}$, with $h = 1, \ldots, 8$, it is possible to cluster them in six groups G_{Aa}, G_{Ab}, ..., G_{Bb}, G_{Bc}. In addition, by clustering the groups that share the measurement of the same phase current, three classes C_{a}, C_{b}, C_{c} are obtained.

The algorithm, by operating a threshold monitoring to each currents' sum $\boldsymbol{\Sigma}_{ifh}$, generates the following diagnostic outputs:

- if all currents' sums included in a *single group* exceed a threshold, a *fault to the sensor* providing the common measurement in the group is detected and isolated;
- if all currents' sums included in a *single class* exceed a threshold, a *fault to both sensors* related to the common phase in the class is detected and isolated;
- if all currents' sums in *all classes* exceed a threshold, a *fault to a motor phase* is detected. In this case, the fault is not directly isolated, since the algorithm does not provide an indication on which phase is failed, but the fault isolation can be obtained by successive ground tests (SBIT, IBIT or MBIT).

4.2.4 *Primary Flight Control Electro-Mechanical Actuator for Medium Altitude Long Endurance Unmanned Aerial Vehicle*

Objectives. The project, developed as a collaboration between the University of Pisa (Italy) and the UAS Division of Leonardo Velivoli (former Alenia SIA, *Società Italiana Avionica*), was part of a more extended program developed by Leonardo Velivoli up to the preliminary design review, aiming to the reliability and performance enhancement of the Flight Control System (FCS) of a Medium Altitude Long Endurance Unmanned Aerial System (MALE UAS). The basic idea of the project was that the major faults of position-controlled flight controls can be detected via *position-tracking monitors*, given that the models provide a satisfactory balance between required accuracy and available computational resources.

The proposed approach, reducing both the software complexity and the number of additional sensors dedicated to monitoring, permits to avoid conflicting indications among monitors and/or uncertainties about corrective actions.

Diagnosis Methods. By applying a model-based approach, the project aimed to:

- the development of *real-time* position-tracking monitors, performing *fault detection, isolation* and *condition monitoring*, to be used as analytical redundancy and implemented in the FCC Continuous Built-in Test (CBIT) in Sect. 4.3.1;
- the development of *prognostic techniques* for the EMA freeplay identification, by using high-fidelity models of the system dynamics (Sect. 4.3.2).

4.2.4.1 Flight Control System Description

In the reference MALE UAS, the actuation of the primary flight controls is obtained by a set of nine EMAs with rotary outputs [15], dedicated to the following aerodynamic movables, see Fig. 4.19:

- four ailerons (two per main wing), for the roll rate control;
- two rudders (one per vertical tail wing), for the yaw rate control;
- three elevators (on horizontal tail wing), for the pitch rate control.

The FCS is designed with a redundant architecture, with three independent Flight Control Computers (FCCs) acting in cross-lane paradigm, so that the system can tolerate one FCC failure without loss of performance.

As shown in Fig. 4.19b, each FCC is connected to each EMA actuator control unit (ACU), which is composed of two computing sections, dedicated to monitoring (MON lane) and closed-loop control (CON lane). Each ACU CON lane is connected to a specific FCC, while the MON lane is signaled by the resting ones. A high-speed cross-lane data link permits to exchange the information between different lanes. The EMA power stage control is performed via a voting-monitoring technique, by using the commands of all FCCs.

Table 4.13 MALE UAS FCS: interfaces between EMA ACU lanes and the FCCs

Flight control function	Movable/EMA	Commands from FCCs to ACU		
		CON lane	MON lane	
Roll	Outboard left aileron	1	2	3
	Outboard right aileron	3	1	2
	Inboard left aileron	2	1	3
	Inboard right aileron	2	1	3
Pitch	Left elevator	1	2	3
	Central elevator	2	1	3
	Right elevator	3	1	2
Yaw	Left rudder	3	1	2
	Right rudder	1	2	3

The interfaces between the FCCs and the EMA ACU sections are reported in Table 4.13 (note that Fig. 4.19b can be referred to outboard left aileron, left elevator or right rudder): it is worth noting that that the FCS is capable of controlling the flight with any combination of two out of three FCCs.

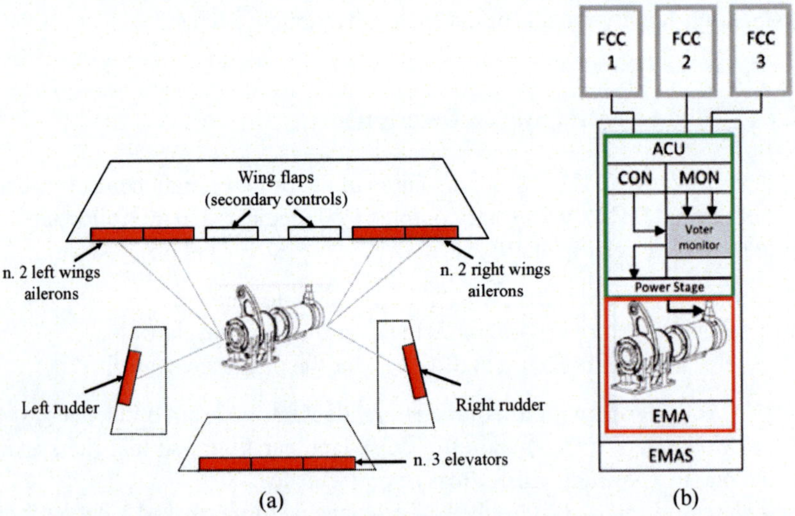

Fig. 4.19 MALE UAS FCS. **a** actuation system layout; **b** EMA interface with FCCs. Reproduced from [16]—originally published open access and licensed under CC-BY 4.0. https://journals.sagepub.com/doi/pdf/10.1177/1687814018768146

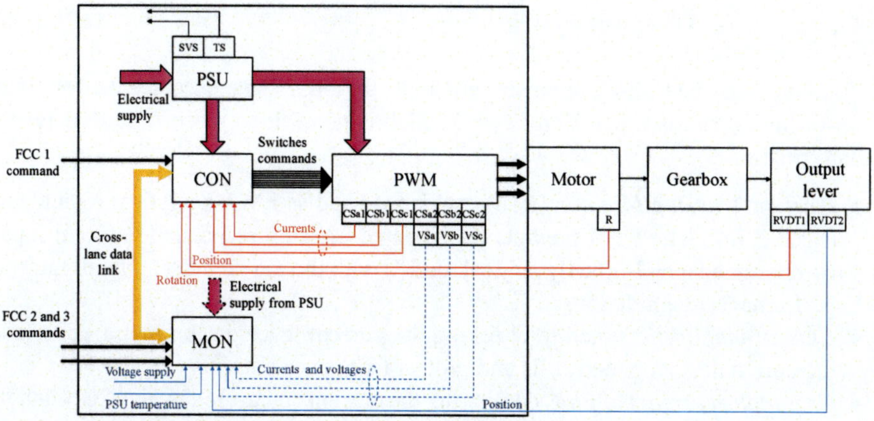

Fig. 4.20 CON-MON ACU of the fault-tolerant EMA. Reproduced from [16]—originally published open access and licensed under CC-BY 4.0. https://journals.sagepub.com/doi/pdf/10.1177/1687814018768146

4.2.4.2 Electro-Mechanical Actuator Description

The flight control EMA is composed of:

- simplex three-phase BLACM, driven via FOC technique;
- phase-isolating power bridge with twelve MOSFET switches, in which each phase is driven via monophase PWM method;
- ECU with CON-MON architecture, implementing the diagnostic algorithms and the closed-loop control functions (based on nested loops on motors' currents, motors' speed, and output position);
- mechanical transmission with epicyclical internal reducer and a four-bar linkage, connecting the output EMA shaft with the aerodynamic surface.

As schematically depicted in Fig. 4.20, the system enables the CON lane to receive feedback from:

- n. 3 current sensors (one per phase);
- n. 1 resolver, measuring the motor angle;
- n. 1 RVDT transducer, measuring the output rotation,

while the MON lane collects data from:

- n. 1 supply voltage sensor;
- n. 1 temperature sensor, installed on the ACU power supply unit (PSU);
- n. 3 current sensors (one per phase);
- n. 3 voltage sensors (one per phase);
- n. 1 RVDT transducer, measuring the output rotation.

4.2.4.3 Fault Diagnosis System

To implement the EMA diagnostic functions, the MON lane executes the following set of fault diagnosis algorithms (with which is possible to perform fault detection and isolation):

- *Position-Tracking Monitor* (PTM), which is used as EMA analytical redundancy (Sect. 2.1.4). The PTM predicts an expected position response to system inputs by relying on a real-time dynamical model, with the aim to detect overall faults or performance degradations;
- *Current monitor*, devoted to checking the current levels in the motor coils, with the aim to detect opened coils and protect from over-currents;
- *Cross-lane current monitor*, comparing the currents measured by the CON and the MON lanes, aimed at detecting sensor faults;
- *In-lane monitors* on RVDT (Rotary Variable Differential Transformers) and resolver, performing consistency checks for the sensors status, related to the detection of components faults;
- *Cross-lane position monitor*, comparing the positions measured by the CON and the MON lanes, to detect transducer faults;
- *Voltage supply monitor*, performing a consistency check on the voltage supply level, to detect a voltage breakdown or a voltage sensor fault;
- *Power Supply Unit (PSU) temperature monitor*, performing an evaluation of the PSU temperature, to detect an abnormal PSU heating or a temperature sensor fault;
- *In-lane CPU monitors*, i.e., watchdogs for both CON and MON lanes;
- *Cross-lane voltage demand monitor*, which performs a comparison between the voltage demands for PWM calculated by the CON and the MON lanes, to detect CPU and I/O faults.

In order to avoid not coherent indications among monitors and/or uncertainties about corrective actions, the detection of EMA faulty behavior and the fail-safe engagement are assigned to the PTM only, while the other monitors are used for the faults' isolation. Starting from this idea, two real-time executable versions of the PTM with different levels of complexity are developed to evaluate the fault detection performances in case of relevant faults (Sect. 4.3.1, [11, 13, 16]).

4.3 Model-Based Approaches

Model-based approaches can be classified, in addition to the specific methodology employed for residuals' generation (see Sect. 3.2.1), also according to the level of complexity of the system models:

- *real-time model-based monitoring* requires the use of reduced-order dynamic models, and they are suitable for CBIT procedures, which are executed throughout the aircraft mission. This approach, necessary when the tolerable fault latencies are very small (i.e., fault diagnosis is expected to be fast), aims at detecting fault modes

characterized by constant failure rate (i.e., random regimes) and abrupt occurrence (e.g., electrical and electronic faults). An example of *real-time model-based* approach is given in Sect. 4.3.1 for *fault diagnosis and condition monitoring*.
- *high-fidelity model-based monitoring* entails the use of accurate dynamic models capable of simulating the system behavior with high level of granularity and resolution [36], and it aims at detecting all the major fault modes of the system. For this reason, this approach can be also suitable for prognosis purposes, by targeting faults characterized by failure rates that increase with the operative time (i.e., wear regimes) and by a slow evolution of the malfunction (e.g., mechanical faults). Examples of *high-fidelity* model-based approaches are given in Sects. 4.3.2 and 4.3.3, with reference to *fault diagnosis and prognosis*, respectively.

In particular, we present:

- a fault detection, isolation, and condition monitoring approach for diagnosis of open and short circuits, motor permanent magnet degradation, and decrease of voltage supply level, based on real-time executable models 4.3.1 (MALE UAS project);
- a prognostic method that deals with freeplay estimation for preventing jamming events, based on high-fidelity models, in Sect. 4.3.2 (MALE UAS project);
- a fault detection, isolation, and condition monitoring approach for jamming diagnosis, based on high-fidelity models 4.3.3 (REPRISE project—phase 2).

4.3.1 Fault Diagnosis via Real-Time Executable Models

Model-based monitoring during flight implies that dynamic models must be real-time executable (i.e., computed within the update sampling rate of the control/monitoring electronics). This requirement can nowadays be satisfied only by reduced models, pursuing a satisfactory balance between prediction accuracy and computational resources [13].

A relevant example is given in [16], in which a real-time model-based approach is applied to the FDI and condition-monitoring of the flight control EMAs of a MALE UAS, by developing two real-time position predictors (PTMs, Sect. 4.2.4) characterized by different levels of complexity. The PTMs are based on time-discrete models that detect a malfunction when the actual position feedback deviates from the prediction for a predefined threshold. Both PTMs receives the same inputs and implements the same Fault Detection Logic (FDL), but they differ for the actuator speed estimation.

4.3.1.1 Fault Detection Logic

The FDL is the same for both versions of the PTM and it is represented by the flow chart in Fig. 4.21. The FDL receives as inputs the normalized error between the

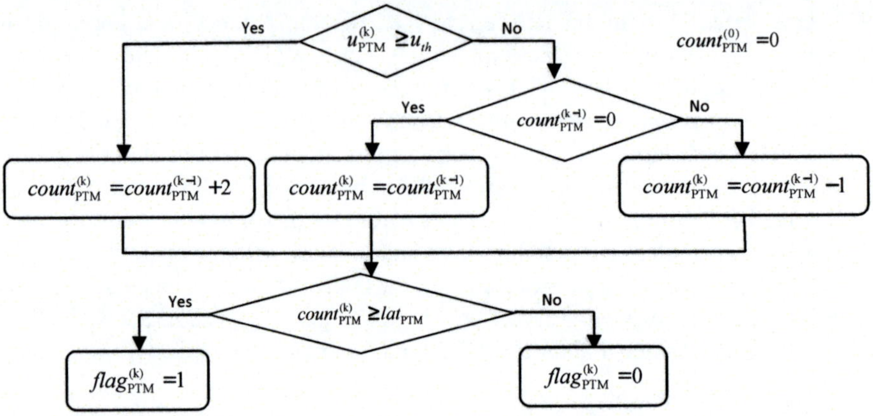

Fig. 4.21 Fault detection logic flow chart. Reproduced from [16]—originally published open access and licensed under CC-BY 4.0. https://journals.sagepub.com/doi/pdf/10.1177/1687814018768146

PTM prediction $\theta_{aPTM}(t)$ and the actual position measured by the MON lane RVDT $\theta_{aML}(t)$. The normalized error $u_{PTM}(t)$ is computed as:

$$u_{PTM}(t) = \frac{|\theta_{aPTM}(t) - \theta_{aML}(t)|}{\theta_{aSAT}}, \tag{4.4}$$

where θ_{aSAT} is the actuator endstroke.

As described in Fig. 4.21, if the normalized error u_{PTM} exceeds a predefined threshold u_{th}, a fault-state counter $count_{PTM}$ is incremented by 2. On the contrary, there are two possibilities: if the counter value was different from zero at the previous step, it is decremented by 1, otherwise it is maintained at zero. If the counter $count_{PTM}$ reaches a threshold lat_{PTM}, the FDL gives in output the boolean signal $flag_{PTM} = 1$ and a malfunction is detected.

4.3.1.2 Real-Time Modeling

Both real-time predictors are based on Eqs. (4.5)–(4.6), which are obtained starting from a 1st-order equivalent monophase model of the BLACM (Sect. 4.3.2.3). Equation (4.5) represents the basic behavior of the motor/gearbox assembly, while (4.6) describes the control laws of the three nested regulators of the system on motor current, motor speed and output rotation.

$$\begin{cases} L\dfrac{di(t)}{dt} = V_c(t) - Ri(t) - K_t\tau_g\dot{\theta}_a(t) \\ (J_m\tau_g^2 + J_a)\,\ddot{\theta}_a(t) = K_t\tau_g i(t) + T_a(t), \end{cases} \tag{4.5}$$

$$\begin{cases} V_c(t) = K_i\big(i_d(t) - i(t)\big) \\ i_d(t) = K_\omega\left(\omega_{md}(t) - \tau_g\dot{\theta}_a(t)\right) \\ \omega_{md} = K_\theta\left(\theta_{ad}(t) - \theta_a(t)\right). \end{cases} \tag{4.6}$$

In (4.5), i is the motor current, θ_a is the actuator output rotation, V_c is the control voltage, T_a is the external torque, τ_g is the gearbox ratio, and J_a is the output inertia, while J_m, L, R and K_t are the motor inertia, inductance, resistance and back-electromotive force coefficient, respectively. In (4.6), K_i, K_ω and K_θ are the gains of the current, speed and deflection proportional regulators respectively, while i_d, ω_{md} and θ_{ad} are the demands of motor current, motor speed, and surface deflection.

In the following, we will drop the dependence of the variables from the time t for easiness of exposition.

First-order predictor (Outer Loop Monitor, OLM). The first version of the PTM is defined as Outer Loop Monitor (OLM). Here, the prediction is obtained by assuming that the speed control loop performs a perfect tracking performance, that is (4.7)

$$\tau_g\dot{\theta}_a = \omega_{md}, \tag{4.7}$$

so that, by substituting (4.7) into (4.6), the actuator output speed is given by

$$\theta_a = p_a\left(\theta_{ad} - \theta_a\right) \tag{4.8}$$

where $1/p_a$ can be interpreted as the time constant of a first-order LTI model approximating the position response (4.9):

$$p_a = \frac{K_\theta}{\tau_g}. \tag{4.9}$$

The OLM uses a time-discrete version of (4.8) and additional conditions to take into account the speed saturation limits, due to the maxima levels of motor voltage and output rotation endstrokes. The result is the *first-order nonlinear predictor* given by (4.10)–(4.12), that operates at T_s sample time (note that z represents the discrete-time operator, i.e., at the k-th sample, $y(k) = z \cdot y(k-1)$):

$$\omega_{a\text{OLM}} = \begin{cases} p_a\left(\theta_{ad} - \theta_{a\text{OLM}}\right) & |\theta_{ad} - \theta_{a\text{OLM}}| < \dfrac{\omega_{a\text{SAT}}}{p_a} \\ \omega_{a\text{SAT}} \cdot \text{sign}\left(\theta_{ad} - \theta_{a\text{OLM}}\right) & |\theta_{ad} - \theta_{a\text{OLM}}| \geq \dfrac{\omega_{a\text{SAT}}}{p_a} \end{cases} \tag{4.10}$$

$$\theta_{a\text{OLM}} = \begin{cases} \dfrac{\omega_{a\text{OLM}}T_s}{z-1} & |z^{-1}\theta_{a\text{OLM}}| < \theta_{a\text{SAT}} \\ \theta_{a\text{SAT}} \cdot \text{sign}\left(z^{-1}\theta_{a\text{OLM}}\right) & |z^{-1}\theta_{a\text{OLM}}| \geq \theta_{a\text{SAT}} \end{cases} \tag{4.11}$$

$$\omega_{a\text{SAT}} = K_{v\omega} V_s \tag{4.12a}$$

$$\theta_{a\text{PTM}} = \theta_{a\text{OLM}} \tag{4.12b}$$

In (4.10)–(4.12), θ_{ad} and V_s are the deflection demand and the supply voltage (PTM inputs), $\omega_{a\text{OLM}}$ is the predicted speed and $\theta_{a\text{OLM}}$ (i.e., $\theta_{a\text{PTM}}$ in (4.4)) is the predicted position. The OLM thus requires four parameters: the actuator endstroke $\theta_{a\text{SAT}}$, the voltage supply to motor speed gain $K_{v\omega}$, actuator speed saturation $\omega_{a\text{SAT}}$, and the quantity p_a defined in (4.9).

Second-order predictor (Inner Loop Monitor, ILM). The second version of the PTM is defined as Inner Loop Monitor (ILM). In this case, the prediction is obtained by assuming that no external torque is applied, i.e., $T_a = 0$ in (4.5), and that the current control loop has ideal tracking behavior, i.e., (4.13)

$$i = i_d = K_\omega \left(\omega_{md} - \tau_g \dot{\theta}_a\right) \tag{4.13}$$

Substituting (4.13) into (4.5), we have

$$\left(J_m \tau_g^2 + J_a\right) \ddot{\theta}_a = K_\omega K_t \tau_g \left(\omega_{md} - \tau_g \dot{\theta}_a\right), \tag{4.14}$$

$$\ddot{\theta}_a = p_v \left(\frac{\omega_{md}}{\tau_g} - \dot{\theta}_a\right), \tag{4.15}$$

where $1/p_v$ can be interpreted as the time constant of a first-order LTI model approximating the actuator speed response, (4.16)

$$p_v = \frac{K_\omega K_t \tau_g^2}{J_m \tau_g^2 + J_s}. \tag{4.16}$$

The actuator output acceleration is obtained by

$$\ddot{\theta}_a = p_v \left[\frac{K_\theta}{\tau_g} \left(\theta_{ad} - \theta_a - \dot{\theta}_a\right)\right] = p_v \left[p_a \left(\theta_{ad} - \theta_a\right) - \dot{\theta}_a\right]. \tag{4.17}$$

The ILM uses a time-discrete version of (4.17), together with saturation conditions for acceleration (due to maxima levels of motor currents), speed and output rotations. The result is the *second-order nonlinear predictor* given by (4.18)–(4.23)

$$\omega_{ad} = p_A \left(\theta_{ad} - \theta_{a\text{ILM}}\right) \tag{4.18}$$

$$\alpha_{a\text{ILM}} = \begin{cases} p_v \left(\omega_{ad} - \omega_{a\text{ILM0}}\right) & \left|\omega_{ai} - \omega_{a\text{ILM0}}\right| < \dfrac{\alpha_{a\text{SAT}}}{p_v} \\ \alpha_{a\text{SAT}} \cdot \text{sign}\left(\omega_{ai} - \omega_{a\text{ILM0}}\right) & \left|\omega_{ai} - \omega_{a\text{ILM0}}\right| \geq \dfrac{\alpha_{a\text{SAT}}}{p_v} \end{cases} \tag{4.19}$$

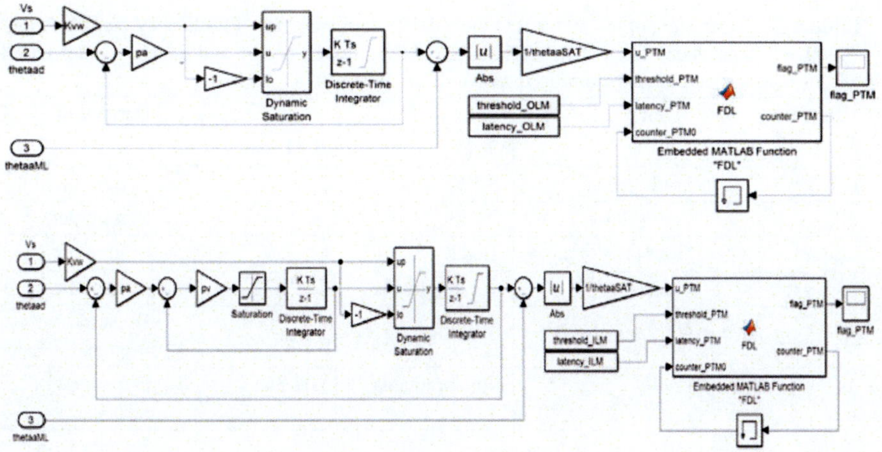

Fig. 4.22 MATLAB-Simulink models of the PTMs: (top) OLM; (bottom) ILM. Reproduced from [16]—originally published open access and licensed under CC-BY 4.0. https://journals.sagepub.com/doi/pdf/10.1177/1687814018768146

$$\omega_{a\text{ILM}0} = \alpha_{a\text{ILM}} \frac{T_s}{z-1} \tag{4.20}$$

$$\omega_{a\text{ILM}} = \begin{cases} \omega_{a\text{ILM}0} & |\omega_{a\text{ILM}0}| < \omega_{a\text{SAT}} \\ \omega_{a\text{SAT}} \cdot \text{sign}\left(\omega_{a\text{ILM}0}\right) & |\omega_{a\text{ILM}0}| \geq \omega_{a\text{SAT}} \end{cases} \tag{4.21}$$

$$\theta_{a\text{ILM}} = \begin{cases} \dfrac{\omega_{a\text{ILM}} T_s}{z-1} & |z^{-1}\theta_{a\text{ILM}}| < \theta_{a\text{SAT}} \\ \omega_{a\text{SAT}} \cdot \text{sign}\left(z^{-1}\theta_{a\text{ILM}}\right) & |z^{-1}\theta_{a\text{ILM}}| \geq \theta_{a\text{SAT}} \end{cases} \tag{4.22}$$

$$\theta_{a\text{PLM}} = \theta_{a\text{ILM}} \tag{4.23}$$

in which $\omega_{a\text{ILM}0}$ is the actuator speed feedback, while $\alpha_{a\text{ILM}}$, $\omega_{a\text{ILM}}$ and $\theta_{a\text{ILM}}$ (i.e., $\theta_{a\text{PTM}}$) are the predicted acceleration, speed, and position, respectively. The ILM predictor is characterized by six parameters: the four parameters of the OLM, plus the quantity p_v and the acceleration saturation $\alpha_{a\text{SAT}}$. The MATLAB-Simulink implementation of the two PTM versions is reported in Fig. 4.22.

4.3.1.3 Definition of the PTMs' Parameters

At first sight, the tuning of the PTMs appears an issue: seven quantities are needed for the OLM (u_{th}, lat_{PTM}, T_s, $\theta_{a\text{PTM}}$, $K_{v\omega}$, $\omega_{a\text{SAT}}$ and p_a), and nine for the ILM (the OLM ones, plus $\alpha_{a\text{SAT}}$ and p_v). Nevertheless, the parameters related to EMA performance limits ($\theta_{a\text{SAT}}$, $K_{v\omega}$, $\omega_{a\text{SAT}}$, and $\alpha_{a\text{SAT}}$) can be directly derived from the system design,

Table 4.14 Parameters of the PTMs. (*) the value is equivalent to a FDI latency ranging from 80 to 160 ms. Reproduced from [16]—originally published open access and licensed under CC-BY 4.0. https://journals.sagepub.com/doi/pdf/10.1177/1687814018768146

Parameter	Unit	Value		Derivation
		OLM	ILM	
θ_{aSAT}	rad	0.524		System data
$K_{v\omega}$	rad/(sV)	0.733		System data
ω_{aSAT}	rad/s	1.047		System data
α_{aSAT}	rad/s^2	Not applicable	121.671	System data
p_a	rad/s	41.469		Dynamic response
p_v	rad/s	Not applicable	207.345	Dynamic response
T_s	ms	4		Requirement
lat_{PTM}	–	40 (*)		Requirement
u_{th}	–	0.004	0.002	Tuning

while dynamic response characteristics (p_a and p_v) can be estimated via experiments. Furthermore, the monitoring sample rate T_s is driven by the real-time execution requirement (in this application, $T_s = 4$ ms), while the fault-counter threshold lat_{PTM} is imposed by the maximum allowable fault latency (in this application FDI latency must be lower than 200 ms).

For these reasons, the parameter tuning for both PTMs was limited to the normalized error threshold u_{th} only, by identifying the value for which no false alarms arise and any mechanical jamming fault is detected within 100 ms. The final PTMs' parameters are reported in Table 4.14.

4.3.1.4 Testing Method and Failure Modes Definition

A high-fidelity EMA model developed in [12, 14] is used to compare the two PTM versions. The model includes the simulation of:

- 3-phases BLACM with FOC technique;
- SVPWM drive of the EMA power stage;
- errors on sensors, such as bias, drift, noise;
- friction, freeplay and first vibrational mode of the mechanical transmission;
- aerodynamic loading;
- major EMA faults (e.g., mechanical jamming, motor faults, sensor faults).

The command time history of a light military jet trainer elevator, during severe pull-up/pull-down maneuvres [11], is used as input to the high-fidelity model and the PTMs. Open or short circuit faults to the motor phases and abrupt voltage decrease, as well as a temperature-induced degradation of the motor magnet properties [21] (a more "hidden" fault difficult to identify with sensors), have been injected in the model to evaluate the PTMs' performance.

This led to five test cases:

1. Normal operation (no faults);
2. Failure mode 1 (FM1), i.e., no faults for $t < 10.4$ s, and
   - first open circuit fault for $10.4\,\text{s} \le t < 20.8$s
   - second open circuit fault for $t \ge 20.8$ s
3. Failure Mode 2 (FM2), i.e., no faults for $t < 10.4$ s, and
   - first short circuit fault for $10.4\,\text{s} \le t < 20.8$ s
   - second short circuit fault for $t \ge 20.8$ s
4. Failure Mode 3 (FM3), i.e., no faults for $t < 10.4$ s, and a step-wise 40% performance degradation of the motor permanent magnet for $t \ge 10.4$ s;
5. Failure Mode 4 (FM4), i.e., no faults for $t < 10.4$ s, and a step-wise 30% decrease of the voltage supply level (with respect to the normal operation) for $t \ge 10.4$ s.

The tests have been also used to verify the applicability of the PTM predictors in case of aerodynamic loading. Since both predictors are developed by referring to the actuator response at zero external load, the accuracy is expected to lower during maneuvres with fast and large-amplitude deflections (i.e., high dynamic loads), due to the actuator dynamic compliance.

4.3.1.5 Fault Diagnosis Performances

The fault diagnosis performances of the developed PTMs are documented in Figs. 4.23, 4.24, 4.25, 4.26, 4.27, 4.28, 4.29, 4.30 and 4.31 (in terms of normalized time-domain responses of output position, motor speed, motor current, aerodynamic load), while the fault detection latencies are reported in Table 4.15.

The results demonstrate that the two versions of the PTM have similar performances in terms of false alarms rejection (Fig. 4.25), as well as for the detection of motor coil faults (both are insensitive to the first fault, while the second one is promptly detected, Fig. 4.25) and for the voltage decrease detection. On the other hand, only the ILM is capable of detecting magnet degradation phenomena (Fig. 4.31). A low sensitivity to actuator loads is present in both predictors.

As shown in Figs. 4.23, 4.24 and 4.25, both algorithms are very robust against false alarms: even when the EMA is demanded to move with high accelerations (e.g., at $t = 1.3$ s, where an abrupt speed change is commanded), the maxima values of the fault counters are lower than 40% of the detection threshold for the OLM, and 30% for the ILM.

The first electrical fault (for both FM1 and FM2 cases) is not detected by either of the algorithms. The effects of the first coil fault on the EMA position response are minor (Fig. 4.26 and 4.27), thanks to the "phase-isolating" design of the EMA power stage (Sect. 4.2.4). On the other hand, both algorithms succeed in detecting the second coil fault (for both FM1 and FM2 cases), with a fault-detection latency of about 1 s for the open circuits and 0.1 s for the short circuits, see Table 4.15.

The results related to the operation with a motor magnet degradation and an abrupt voltage decrease (FM3 and FM4) demonstrate that both algorithms behave satisfactorily with reference to the FM4 (fault latency is small and similar for both PTMs, Table 4.15), while only the ILM is capable of detecting the magnet degradation (Fig. 4.31 and Table 4.15).

Finally, notice how the proposed approach can be used to perform not only fault detection and isolation, but also condition monitoring, by using the fault counter $count_{PTM}$ as indication for the state of system degradation (Figs. 4.25, 4.28, and 4.31).

4.3.2 *Fault Prognosis via High-Fidelity Dynamic Models*

As previously stated, the mechanical transmission jamming is surely the most feared fault in EMAs. Mechanical jamming occurs because the load is transmitted through mechanical contacts with high local stresses, which cause fatigue in the materials. The degradation of the contact surfaces initially implies lower efficiency (with impact on power consumption) and increased freeplay (which, in flight control applications,

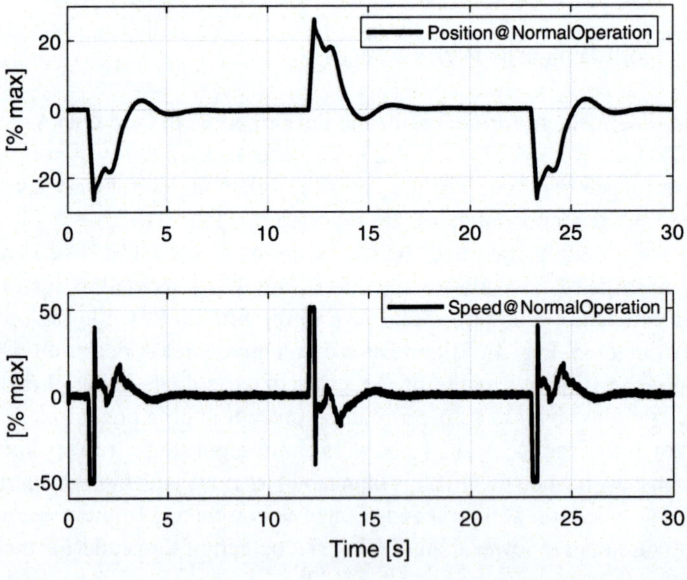

Fig. 4.23 EMA response in normal operation: output position and motor speed. Reproduced from [16]—originally published open access and licensed under CC-BY 4.0. https://journals.sagepub.com/doi/pdf/10.1177/1687814018768146

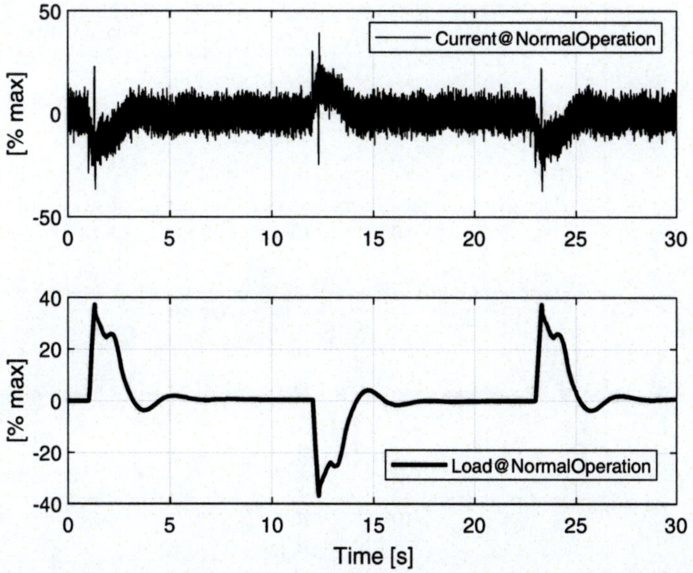

Fig. 4.24 EMA response in normal operation: motor current and aerodynamic load. Reproduced from [16]—originally published open access and licensed under CC-BY 4.0. https://journals.sagepub.com/doi/pdf/10.1177/1687814018768146

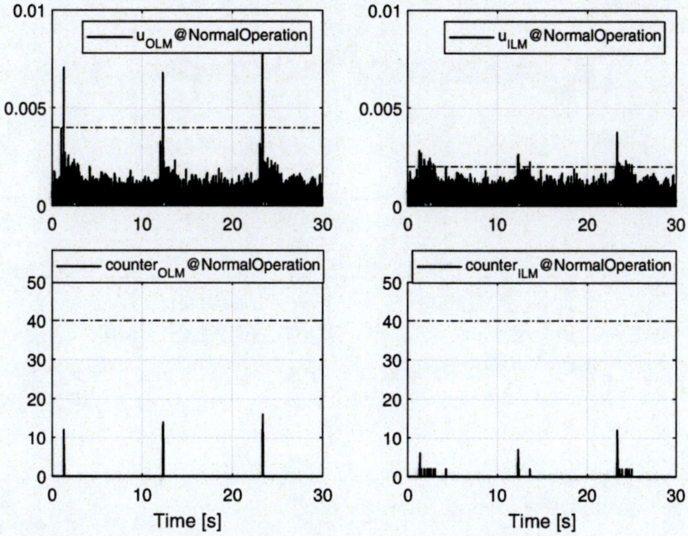

Fig. 4.25 Residuals and fault-state counters of the PTMs' in normal operation. Reproduced from [16]—originally published open access and licensed under CC-BY 4.0. https://journals.sagepub.com/doi/pdf/10.1177/1687814018768146

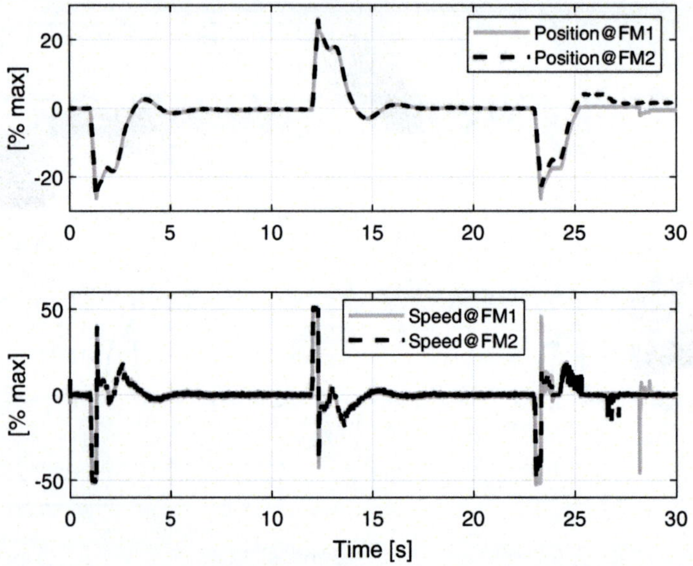

Fig. 4.26 EMA response in FM1 and FM2: output position and motor speed (1st fault at 10.4 s; 2nd fault at 20.8 s). Reproduced from [16]—originally published open access and licensed under CC-BY 4.0. https://journals.sagepub.com/doi/pdf/10.1177/1687814018768146

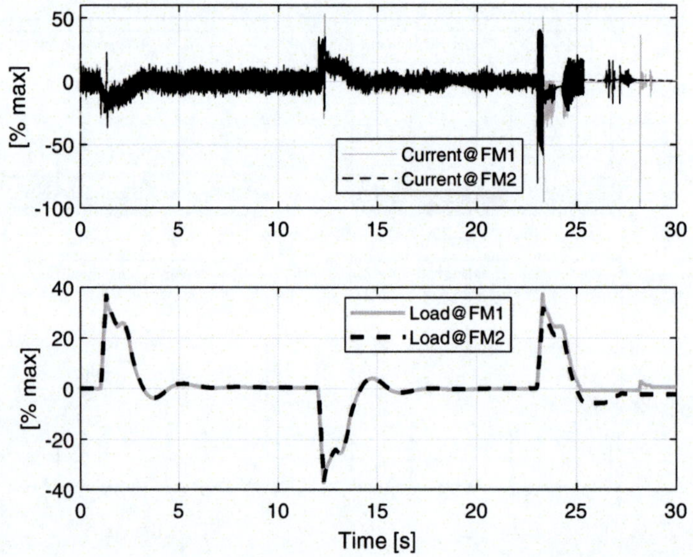

Fig. 4.27 EMA response in FM1 and FM2: motor current and aerodynamic load (1st fault at 10.4 s; 2nd fault at 20.8 s). Reproduced from [16]—originally published open access and licensed under CC-BY 4.0. https://journals.sagepub.com/doi/pdf/10.1177/1687814018768146

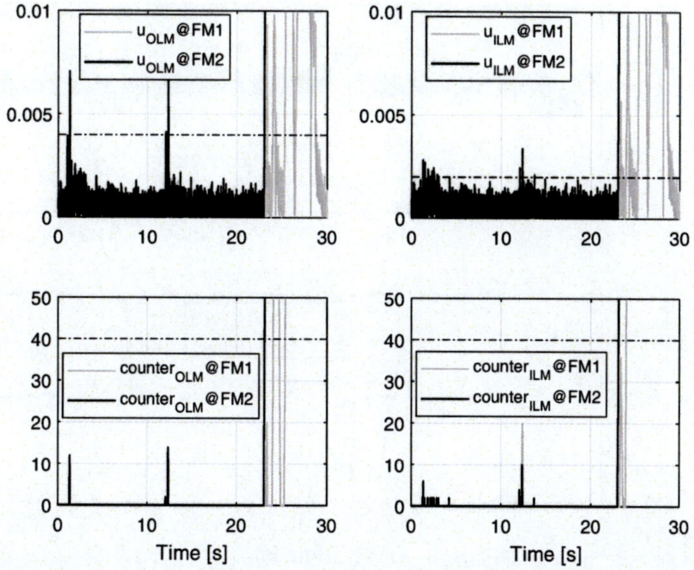

Fig. 4.28 Residuals and fault-state counters of the PTMs in FM1 and FM2 (1st fault at 10.4 s; 2nd fault at 20.8 s). Reproduced from [16]—originally published open access and licensed under CC-BY 4.0. https://journals.sagepub.com/doi/pdf/10.1177/1687814018768146

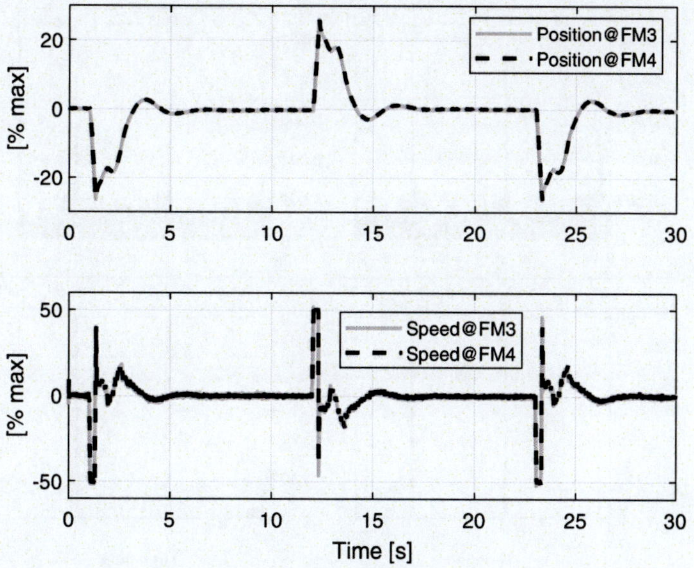

Fig. 4.29 EMA response in FM3 and FM4: output position and motor speed (fault at 10.4 s). Reproduced from [16]—originally published open access and licensed under CC-BY 4.0. https://journals.sagepub.com/doi/pdf/10.1177/1687814018768146

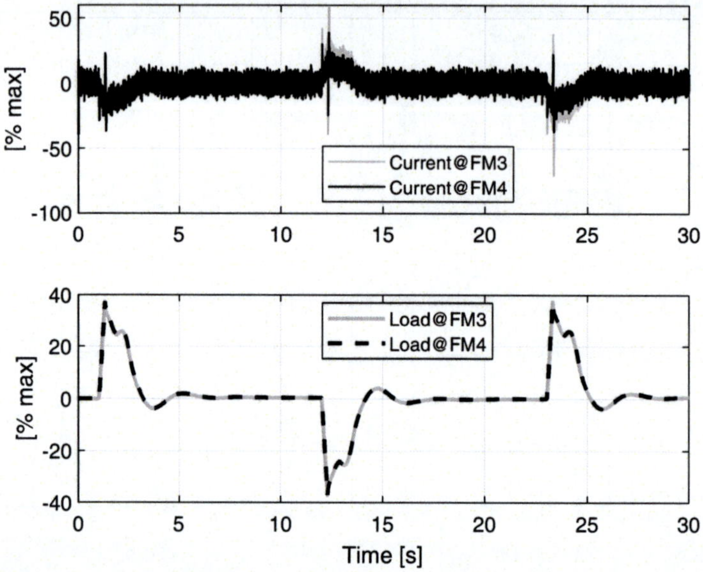

Fig. 4.30 EMA response in FM3 and FM4: motor current and aerodynamic load (fault at 10.4 s). Reproduced from [16]—originally published open access and licensed under CC-BY 4.0. https://journals.sagepub.com/doi/pdf/10.1177/1687814018768146

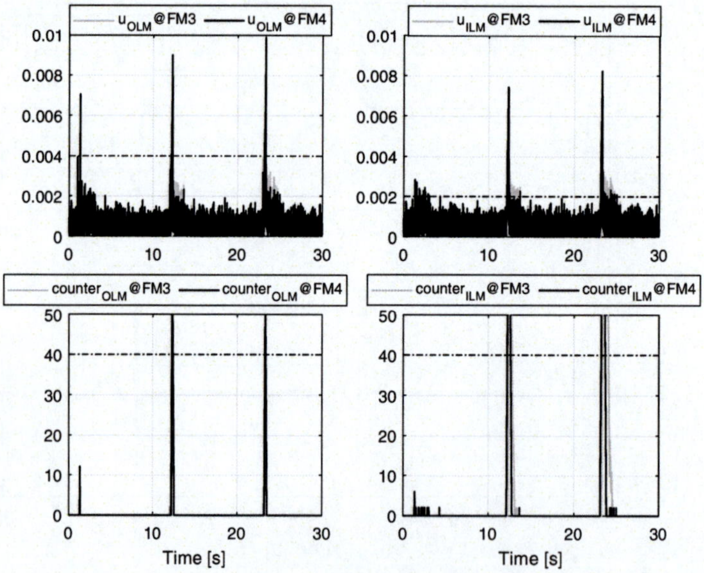

Fig. 4.31 Residuals and fault-state counters of the PTMs in FM3 and FM4 (fault at 10.4 s). Reproduced from [16]—originally published open access and licensed under CC-BY 4.0. https://journals.sagepub.com/doi/pdf/10.1177/1687814018768146

Table 4.15 Report of PTMs' latencies (evaluated from the beginning of the maneuvre next to fault injection). Reproduced from [16]—originally published open access and licensed under CC-BY 4.0. https://journals.sagepub.com/doi/pdf/10.1177/1687814018768146

	FM1 (open circuits)		FM2 (short circuits)		FM3 (degraded magnet)	FM4 (30% voltage decrease)
	1st	2nd	1st	2nd		
OLM	Not detected	1.04 s	Not detected	0.1 s	Not detected	0.29 s
ILM	Not detected	1 s	Not detected	0.1 s	0.41 s	0.20 s

can induce dangerous aeroservoelastic interactions). Finally, the degradation can lead to the mechanical block.

The jamming of a safety-critical actuator can represent a failure with catastrophic consequences, which often makes useless architectures with parallel or grouped actuators. Depending on the EMA architecture, several components can cause a jamming fault: gear trains and screw-nut assemblies, primary bearings supporting the actuator loads, secondary bearings included in the motor and the reducer. To avoid jamming, two strategies are typically investigated. The first one is based on jamming isolation inside the actuator (by adding a redundant mechanical channel or by integrating an unlocking device), while the second strategy aims at the fault anticipation, by using Prognostic Condition-Management (PCM) algorithms capable of predicting the actuator Remaining Useful Life (RUL).

The main advantages gained by PCM algorithms are the weight and size reduction, the increase of system reliability, the maintenance optimization (thanks to the minimization of the corrective actions during the service life). As drawbacks, PCM algorithms typically require relevant computing resources and additional sensors. In addition, the design and the validation of the PCM algorithms can result in prohibitive costs: actuator nonlinearities, sensor disturbances, and sensitivity to environment and loads can have a strong impact on PCM results, so an in-depth knowledge of the actuator dynamics in both normal and degraded condition is required to develop the algorithms. This can be achieved via experimental activities (*data-driven* approach), in which the actuator response is characterized with respect to several degraded conditions, up to create a reference prognostic database [59].

The resulting rigging costs range from relevant to excessive. Alternatively, when the correlation between the degradation to be identified and the effect to be measured is well established, a model-based approach is preferable. In a *model-based* PCM approach, the prognostic database is made of high-fidelity models capable of simulating the degraded behavior by physical first principles [31, 62]. Nonetheless, at least for the system in healthy, the use of experimental data may be necessary to align the model with the true system (e.g., for estimating unknown model parameters via system identification methods).

The use of PCM algorithms for the freeplay identification can speed up the EMA application in flight controls, since they can anticipate the jamming occurrence and provide protection from aeroservoelastic instability (freeplay on flight surfaces can cause the decrease of flutter speed). Actually, though the EMA freeplay can be minimized with pre-loaded mechanical transmissions, this solution is not suitable for primary flight controls, since they require high dynamic performances. For this reason, EMA for flight controls are prone to freeplay, and their jamming is often anticipated by freeplay increase.

A relevant example of model-based PCM oriented to the freeplay identification is given in [17], where the study was referred to the EMA for MALE UAS flight controls described in Sect. 4.2.4.

4.3.2.1 High-Fidelity Model Features

To design the PCM algorithms, a detailed nonlinear model of the EMA dynamics is developed and validated with experiments in the normal operative condition (i.e., new actuator, no faults). The model includes the simulation of:

- three-phase BLACM, driven via FOC technique;
- phase-isolating power bridge with monophase PWM drive;
- nested digital loops on motor currents, motor speed, and output rotation;
- sensors' errors (bias, noise) and control nonlinearities (saturation, rate limiting, quantization);
- 4-DoF mechanical transmission with equations of motions related to motor, gearbox, output lever, and control surface rotations, including freeplay and sliding friction;
- four-bar linkage kinematics.

4.3.2.2 Model of the Three-Phase Brushless AC Motor

The BLACM model is developed with reference to the schematics shown in Fig. 4.32 (where only one pole pair is depicted to simplify the sketch), under the following basic assumptions:

1. the magnetic nonlinearities of ferromagnetic parts (i.e., hysteresis, saturation) are negligible;
2. the motor is magnetically symmetric with respect to its phases;
3. the permanent magnets are made of rare-earth materials, and the magnet reluctance along the q-axis is infinite with respect to the one along the d-axis;
4. the magnetic coupling of the phases is negligible;
5. the reluctances of the ferromagnetic parts are negligible;
6. the magnetic flux dispersions (i.e., secondary magnetic paths) are negligible;
7. the current drive is operated via FOC technique.

Clarke-Parke transformations (FOC technique). When three-phase Permanent Magnets Synchronous Motors (PMSMs) are controlled via the so-called FOC technique, the analysis of the voltages, currents, and fluxes are made in terms of complex space vectors. In particular, considering the geometrical arrangement of the phases (circumferentially distributed along the 360° electrical cycle), all physical quantities related to a phase evolve with 120° electrical angle lead or delay with respect the others.

Given a phase-related vector in the stator frame $\mathbf{w}_{abc}(t) = [w_a(t), w_b(t), w_c(t)]^\top$, the FOC technique allows to transform it into two reference frames, Fig. 4.33:

- the Clarke frame, a fixed frame (α, β, γ) where the α axis is aligned with phase a;
- the Park frame, a rotating frame (d, q, z) in which the d axis is aligned with the direct axis of the permanent magnet of the rotor.

The Clarke transformation transforms the stator-referenced vector $\mathbf{w}_{abc}$ into the Clarke frame

$$\begin{bmatrix} w_\alpha \\ w_\beta \\ w_\gamma \end{bmatrix} = \mathbf{T}_C \cdot \begin{bmatrix} w_a \\ w_b \\ w_c \end{bmatrix}, \tag{4.24}$$

where

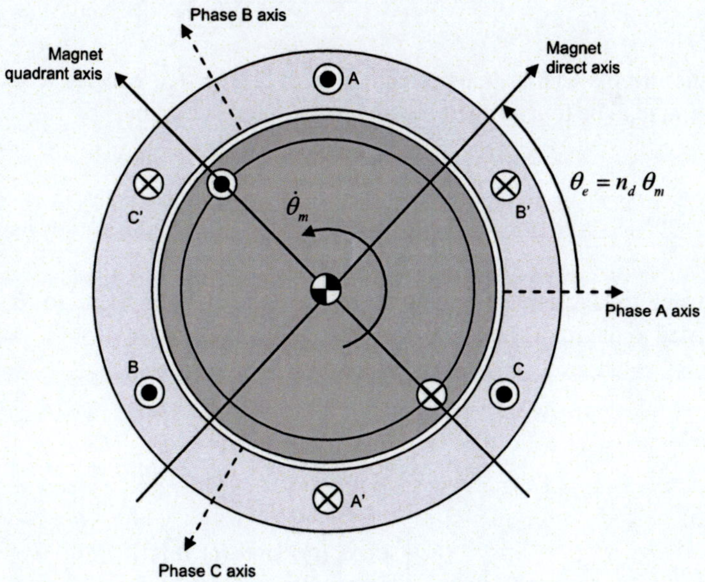

Fig. 4.32 Three-phase permanent magnet synchronous motor schematics (one pole pair)

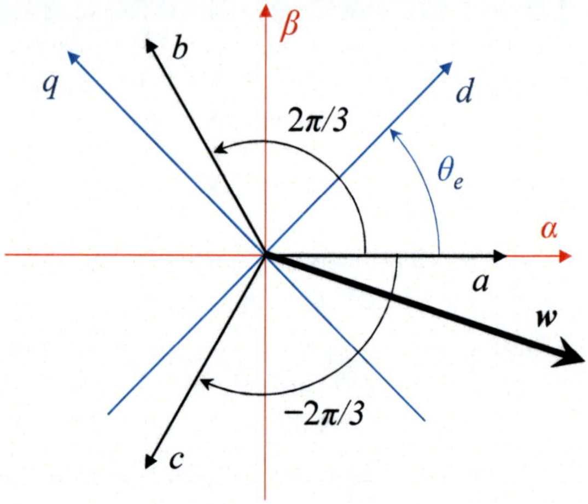

Fig. 4.33 Stator, Clarke, and Park reference frames

$$\mathbf{T}_C = \sqrt{\frac{2}{3}} \cdot \begin{bmatrix} 1 & -\frac{1}{2} & -\frac{1}{2} \\ 0 & \frac{\sqrt{3}}{2} & -\frac{\sqrt{3}}{2} \\ \frac{\sqrt{2}}{2} & \frac{\sqrt{2}}{2} & \frac{\sqrt{2}}{2} \end{bmatrix}, \tag{4.25}$$

while the inverse Clarke transformation, with $\mathbf{T}_C^{-1} = \mathbf{T}_C^{\top}$, enables a vector in the Clarke frame to be transformed into the stator reference frame

$$\begin{bmatrix} w_a \\ w_b \\ w_c \end{bmatrix} = \mathbf{T}_C^{-1} \cdot \begin{bmatrix} w_\alpha \\ w_\beta \\ w_\gamma \end{bmatrix}, \tag{4.26}$$

By using the Park transformation, a vector in the Clarke frame (α, β, γ) can be transported to the rotor frame (d, q, z), so that

$$\begin{bmatrix} w_d \\ w_q \\ w_z \end{bmatrix} = \mathbf{T}_P(\theta_e) \cdot \begin{bmatrix} w_\alpha \\ w_\beta \\ w_\gamma \end{bmatrix}, \tag{4.27}$$

in which

$$\mathbf{T}_P(\theta_e) = \begin{bmatrix} \cos\theta_e(t) & \sin\theta_e(t) & 0 \\ -\sin\theta_e(t) & \cos\theta_e(t) & 0 \\ 0 & 0 & 1 \end{bmatrix}, \tag{4.28}$$

where $\theta_e(t)$ is the electrical angle, related to the motor rotation $\theta_m(t)$ via the number of pole pairs n_d as

$$\theta_e(t) = n_d \cdot \theta_m(t) \tag{4.29}$$

The inverse Park transformation, with $\mathbf{T}_P^{-1}(\theta_e) = \mathbf{T}_P^\top(\theta_e)$ is thus

$$\begin{bmatrix} w_\alpha \\ w_\beta \\ w_\gamma \end{bmatrix} = \mathbf{T}_P^{-1}(\theta_e) \cdot \begin{bmatrix} w_d \\ w_q \\ w_z \end{bmatrix}, \tag{4.30}$$

The combined Clarke-Park transformation finally allows to transform a stator-referenced vector $\mathbf{w}_{abc}$ into a rotor-referenced vector $\mathbf{w}_{dqz}$ (and inversely), so that

$$\begin{bmatrix} w_d \\ w_q \\ w_z \end{bmatrix} = \mathbf{T}_P(\theta_e)\mathbf{T}_C \cdot \begin{bmatrix} w_a \\ w_b \\ w_c \end{bmatrix} = \mathbf{T}_{PC}(\theta_e) \begin{bmatrix} w_a \\ w_b \\ w_c \end{bmatrix}, \tag{4.31}$$

where, by dropping for notational simplicity the dependency of θ_e by the time t

$$\mathbf{T}_{PC}(\theta_e) = \sqrt{\frac{2}{3}} \cdot \begin{bmatrix} \cos\theta_e & \cos\left(\theta_e - \frac{2\pi}{3}\right) & \cos\left(\theta_e + \frac{2\pi}{3}\right) \\ -\sin\theta_e & -\sin\left(\theta_e - \frac{2\pi}{3}\right) & -\sin\left(\theta_e + \frac{2\pi}{3}\right) \\ \frac{\sqrt{2}}{2} & \frac{\sqrt{2}}{2} & \frac{\sqrt{2}}{2} \end{bmatrix}, \tag{4.32}$$

as well as, with $\mathbf{T}_{PC}^{-1}(\theta_e) = \mathbf{T}_{PC}^\top(\theta_e)$

$$\begin{bmatrix} w_a \\ w_b \\ w_c \end{bmatrix} = \mathbf{T}_{PC}^{-1}(\theta_e) \cdot \begin{bmatrix} w_d \\ w_q \\ w_z \end{bmatrix}, \tag{4.33}$$

Magnetic reluctances and magnetic fluxes calculation. The three-phase PMSMs are basically characterized by four magnetic fluxes :

- one linked to the rotor magnet $\varphi_m(t)$;
- three linked to the stator phases $\varphi_a(t)$, $\varphi_b(t)$ and $\varphi_c(t)$,

and four magnetomotive forces

- one induced by the magnet Φ_m;
- three ones due to the current circulation into the stator phases, $Ni_a(t)$, $Ni_b(t)$, and $Ni_c(t)$).

In what follows, we drop the dependence of the variables on time t for ease of explanation.

The fluxes linked to the x-th phase ($x = a, b, c$) and the magnetomotive forces generated by the y-th motor part ($y = a, b, c, m$) are related via the Hopkinson law

$$\begin{bmatrix} \varphi_a \\ \varphi_b \\ \varphi_c \end{bmatrix} = \begin{bmatrix} 1/\mathscr{R}_{aa} & 1/\mathscr{R}_{ab} & 1/\mathscr{R}_{ac} & 1/\mathscr{R}_{am} \\ 1/\mathscr{R}_{ba} & 1/\mathscr{R}_{bb} & 1/\mathscr{R}_{bc} & 1/\mathscr{R}_{bm} \\ 1/\mathscr{R}_{ca} & 1/\mathscr{R}_{cb} & 1/\mathscr{R}_{cc} & 1/\mathscr{R}_{cm} \\ 1/\mathscr{R}_{ma} & 1/\mathscr{R}_{mb} & 1/\mathscr{R}_{mc} & 1/\mathscr{R}_{mm} \end{bmatrix} \cdot \begin{bmatrix} Ni_a \\ Ni_b \\ Ni_c \\ \Phi_m \end{bmatrix}, \tag{4.34}$$

where $\mathscr{R}_{xy}$ is the magnetic reluctance (so $1/\mathscr{R}_{xy}$ is the permeance) due to the circulation of the x-th magnetic flux when the y-th magnetomotive force is applied. The self-permeance of the magnet is given by

$$\frac{1}{\mathscr{R}_{mm}} \approx \frac{1}{\mathscr{R}_m^{(d)} + 2\mathscr{R}_g} = \frac{1}{\mathscr{R}_{m0}}, \tag{4.35}$$

where $\mathscr{R}_m^{(d)}$ is the magnet reluctance along its direct axis and $\mathscr{R}_g$ is the reluctance of the air gap between the stator and the rotor, and $\mathscr{R}_{m0}$ is the total reluctance of the magnet.

The permeance relating the phase fluxes with the permanent magnets are instead given by (4.36)

$$\frac{1}{\mathscr{R}_{am}} = \frac{1}{\mathscr{R}_{m0}}\cos\left(\theta_e\right) \tag{4.36a}$$

$$\frac{1}{\mathscr{R}_{bm}} = \frac{1}{\mathscr{R}_{m0}}\cos\left(\theta_e - \frac{2\pi}{3}\right) \tag{4.36b}$$

$$\frac{1}{\mathscr{R}_{cm}} = \frac{1}{\mathscr{R}_{m0}}\cos\left(\theta_e + \frac{2\pi}{3}\right) \tag{4.36c}$$

Concerning the magnetic reluctances of the phases related to current-induced magnetomotive forces, they are calculated by neglecting their variations with respect to motor angle. For example, the self-reluctance of the phase a is generally expressed by

$$\frac{1}{\mathscr{R}_{aa}} = \frac{1}{\mathscr{R}_{m0}}\cos^2\theta_e + \frac{1}{\mathscr{R}_m^{(q)} + 2\mathscr{R}_g}\sin^2\theta_e, \tag{4.37}$$

where $\mathscr{R}_m^{(q)}$ is the reluctance of the magnet along its quadrant axis.

If we reformulate (4.37) as

$$\frac{1}{\mathscr{R}_{aa}} = \frac{1}{\mathscr{R}_{m0}}\left(\cos^2\theta_e + \frac{\mathscr{R}_{m0}}{\mathscr{R}_m^{(q)} + 2\mathscr{R}_g}\sin^2\theta_e\right), \tag{4.38}$$

and we assume $\mathscr{R}_m^{(q)} \gg \mathscr{R}_{m0}$, we have

$$\frac{1}{\mathscr{R}_{aa}} \approx \frac{1}{\mathscr{R}_{m0}}\cos^2\theta_e = \frac{1}{2\mathscr{R}_{m0}}\left(1 + \cos(2\theta_e)\right), \tag{4.39}$$

where the term related to motor angle induces minor effects on system dynamics, so that (the independence from motor angle implies that all phases have the same self-permeance)

$$\frac{1}{\mathscr{R}_{aa}} = \frac{1}{\mathscr{R}_{bb}} = \frac{1}{\mathscr{R}_{cc}} \approx \frac{1}{2\mathscr{R}_{m0}}. \tag{4.40}$$

Finally, by assuming that the permeance matrix in (4.34) is symmetrical and neglecting the mutual couplings among phases, we obtain

$$\begin{bmatrix} \varphi_a \\ \varphi_b \\ \varphi_c \end{bmatrix} = \frac{1}{\mathscr{R}_{m0}} \cdot \begin{bmatrix} 1/2 & 0 & 0 & \cos\theta_e \\ 0 & 1/2 & 0 & \cos\left(\theta_e - \frac{2\pi}{3}\right) \\ 0 & 0 & 1/2 & \cos\left(\theta_e + \frac{2\pi}{3}\right) \\ \cos\theta_e & \cos\left(\theta_e - \frac{2\pi}{3}\right) & \cos\left(\theta_e + \frac{2\pi}{3}\right) & 1 \end{bmatrix} \cdot \begin{bmatrix} Ni_a \\ Ni_b \\ Ni_c \\ \Phi_m \end{bmatrix}, \tag{4.41}$$

Phase currents' dynamics and motor torque. Having obtained the magnetic fluxes, the dynamic of the phase currents are given by (4.42)

$$V_a = Ri_a + L\frac{di_a}{dt} - \lambda_m n_d \dot{\theta}_m \sin\left(n_d\theta_m\right), \tag{4.42a}$$

$$V_b = Ri_b + L\frac{di_b}{dt} - \lambda_m n_d \dot{\theta}_m \sin\left(n_d\theta_m - \frac{2\pi}{3}\right), \tag{4.42b}$$

$$V_c = Ri_c + L\frac{di_c}{dt} - \lambda_m n_d \dot{\theta}_m \sin\left(n_d\theta_m + \frac{2\pi}{3}\right), \tag{4.42c}$$

where L is the self-inductance of the phases and λ_m is the phase flux linkage due to the rotor magnet

$$L = \frac{N^2}{2\mathscr{R}_{m0}}, \tag{4.43a}$$

$$\lambda_m = \frac{N\Phi_m}{\mathscr{R}_{m0}}. \tag{4.43b}$$

Finally, the motor torque is given by

$$T_m = \lambda_m \left[i_a \sin\left(n_d\theta_m\right) + i_b \sin\left(n_d\theta_m - \frac{2\pi}{3}\right) + i_c \sin\left(n_d\theta_m + \frac{2\pi}{3}\right)\right]. \tag{4.44}$$

It is worth noting that the motor torque can be expressed in the rotor reference frame (d, q, z) by applying the Clarke-Park transformation to the phase currents. By imposing $i_z = 0$, we obtain

$$T_m = k_t i_q, \tag{4.45}$$

where k_t is the torque constant of the BLACM

$$k_t = \sqrt{\frac{3}{2}}\lambda_m n_d \tag{4.46}$$

4.3.2.3 Reduced-Order Brushless AC Motor Models

In the preliminary design phases, the dynamics of BLACMs can be effectively described by a reduced-order model referred to the (d, q, z) frame. Actually, the electrical equations in the stator frame can be written as

$$\mathbf{v}_{abc} = \mathbf{R} \cdot \mathbf{i}_{abc} + \mathbf{L} \cdot \frac{d}{dt}\mathbf{i}_{abc} + \mathbf{e}_{abc}, \tag{4.47}$$

where $\mathbf{v}_{abc} = [V_a \quad V_b \quad V_c]^\top$ is the applied voltages vector, $\mathbf{i}_{abc} = [i_a \quad i_b \quad i_c]^\top$ is the phase currents vector, $\mathbf{R} = R \cdot I_3$, $\mathbf{L} = L \cdot I_3$, $I_3 \in \mathbb{R}^{3\times3}$ is the identity matrix, and $\mathbf{e}_{abc}$ is the back-electromotive forces vector:

$$\mathbf{e}_{abc} = \lambda_m \dot{\theta}_e \cdot \left[\sin\theta_e \quad \sin\left(\theta_e - \frac{2\pi}{3}\right) \quad \sin\left(\theta_e + \frac{2\pi}{3}\right)\right]^\top \tag{4.48}$$

Equation (4.47) can be transformed via the Clarke-Park transform

$$\mathbf{T}_{PC}^{-1} \cdot \mathbf{v}_{dqz} = \mathbf{R}\mathbf{T}_{PC}^{-1} \cdot \mathbf{i}_{dqz} + \mathbf{L}\frac{d}{dt}\left(\mathbf{T}_{PC}^{-1} \cdot \mathbf{i}_{dqz}\right) + \mathbf{T}_{PC}^{-1} \cdot \mathbf{e}_{dqz}, \tag{4.49}$$

obtaining the electrical equations in the (d, q, z) frame

$$\mathbf{v}_{dqz} = \mathbf{R}\mathbf{i}_{dqz} + \mathbf{T}_{PC}\mathbf{L}\frac{d}{dt}\left(\mathbf{T}_{PC}^\top\right)\mathbf{i}_{dqz} + \mathbf{L}\frac{d}{dt}\mathbf{i}_{dqz} + \mathbf{e}_{dqz}. \tag{4.50}$$

The resulting third-order dynamic model in (4.51) is the so-called *equivalent DQZ model* of a BLACM

$$V_d = Ri_d + L\frac{di_d}{dt} - \frac{k_t L}{\lambda_m}\dot{\theta}_m i_q, \tag{4.51a}$$

$$V_q = Ri_q + L\frac{di_q}{dt} - \frac{k_t L}{\lambda_m}\dot{\theta}_m i_d + k_t\dot{\theta}_m, \tag{4.51b}$$

$$V_z = Ri_z + L\frac{di_z}{dt}, \tag{4.51c}$$

where it can be noted that the current dynamics on z-axis is decoupled from the others, and it is typically ignored, so that the equivalent DQZ model is often given as a second-order system.

An additional model reduction can be obtained by assuming that, in (4.51), the direct current i_d is null (i.e., the reference value of the FOC technique), so that we can also define a first-order equivalent monophase model

$$V_q \approx Ri_q + L\frac{di_q}{dt} + k_t\dot{\theta}_m. \tag{4.52}$$

4.3.2.4 Model of the Mechanical Transmission with Freeplay

The prognostic database for the PCM algorithms is made of EMA models characterized by different values of freeplay, and the freeplay effect on the mechanical transmission is simulated by means of a lumped-parameters approach [44, 45, 58, 63].

The torque (and motion) transmission between two rotating bodies with angular freeplay ε is obtained by (4.53)

$$J_1\ddot{\theta}_1 = T_{s\to 1} + \sum_j T_{j1}, \tag{4.53a}$$

$$J_2\ddot{\theta}_2 = T_{s\to 2} + \sum_k T_{k2}, \tag{4.53b}$$

$$T_{s\to 2} = -T_{s\to 1} = -K_s\theta_s - C_s\dot{\theta}_s, \tag{4.53c}$$

where θ_1, θ_2 and J_1, J_2 are the rotation and the moment of inertia of the bodies 1 and 2, respectively; $T_{s\to 1}$, $T_{s\to 2}$ and T_{j1}, T_{k1} are the transmitted torques and the j-th and k-th external torque contributions applied to the body 1 and 2, respectively; θ_s and $\dot{\theta}_s$ are the structural deformation and deformation rate, while K_s and C_s are the torsional stiffness and damping of the mechanical transmission.

The structural deformation is then governed by dynamic equations with two states:

- *state 1 (free motion)*, in which the rotations imply that no contact is established, i.e., $|\theta_2 - \theta_1| < \varepsilon$, and
$$\dot{\theta}_s = -\frac{K_s}{C_s}\theta_s; \tag{4.54}$$
- *state 2 (connection)*, in which the rotations permit to establish the contact, i.e., $|\theta_2 - \theta_1| \geq \varepsilon$, and

$$\dot{\theta}_s = \dot{\theta}_2 - \dot{\theta}_1, \tag{4.55a}$$

$$\theta_s = \theta_2 - \theta_1 - \varepsilon \cdot \text{sign}\,(\theta_2 - \theta_1)\,. \tag{4.55b}$$

The above model is applied at two locations in the mechanical transmission train. The first one is internal to the actuator, i.e., between the gearbox output θ_{GB} and the output lever θ_{out} (in (4.53a)–(4.55), $\theta_1 = \theta_{GB}$ and $\theta_2 = \theta_{out}$, Fig. 4.34); while the other is external, i.e., between the output lever and the control surface (in (4.53a)–(4.55), $\theta_1 = \theta_{out}$ and $\theta_2 = \theta_{cs}$, Figs. 4.35 and 4.34). This allows analyzing the effects of freeplay on EMA dynamics when the mechanical degradation is both "inside" and "outside" the position control loop.

4.3.2.5 Fault Prognosis Algorithm

The basic idea of the developed Prognostic and Condition Monitoring (PCM) algorithm is that the freeplay, in position-controlled EMAs, implies the onset of limit cycle oscillations, with amplitude and frequency that depend on the freeplay size, which can be related to the EMA age and Residual Useful Life (RUL). The PCM

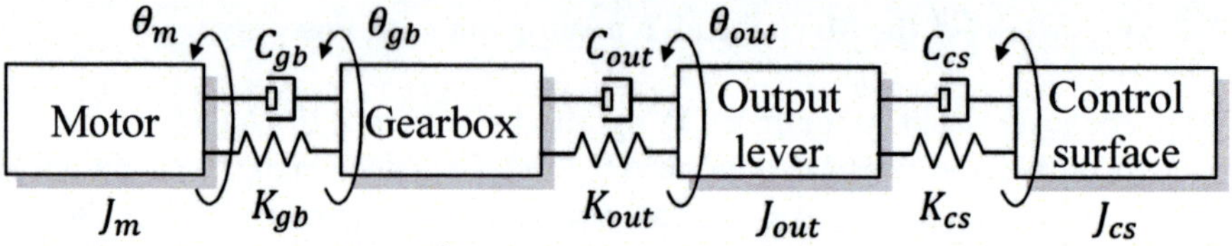

Fig. 4.34 Schematics of the 4 DoF-model of the mechanical transmission

Fig. 4.35 Four-bar linkage connecting the output lever with the aerodynamic surface

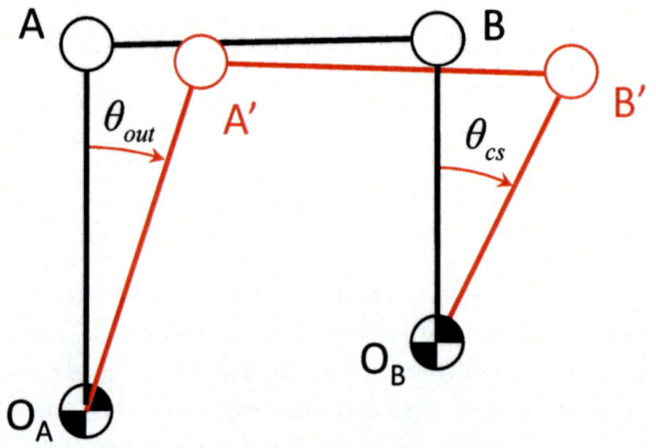

algorithm, designed to be used during the Maintenance Built-in Tests (MBITs), collects the time histories of the EMA sensors' during position-tracking tests capable of inducing freeplay-related limit cycles, and it operates a signals' treatment (FFT, normalization, amplification, filtering) aiming to valorize the limit cycle content with respect to both high-frequency disturbances and low-frequency dynamics.

The MBIT is thus simulated on each prognostic model, and the related limit cycles are characterized, weighed against uncertainties (sensors' errors and parameters' variations), and collected to create a PCM database. The amplitude and the frequency of the measured limit cycle oscillations are finally compared with the database, by generating a freeplay estimation and a RUL prediction.

The PCM design is based on the assumption that periodical MBITs are performed on the flight control EMA during its service life, and that these tests consist of commanding the system with a large-amplitude square-wave demand at low frequency (i.e., an input that is capable of inducing the onset of the freeplay-related limit cycles). As an example, the time histories of the EMA motor angle θ_m during two MBITs with different values of freeplay are reported in Fig. 4.36, where the onset of limit cycles is clearly observable (the reported results are referred to an equal allocation of freeplay in the two locations of the mechanical transmission, i.e., the two freeplays cause the same output backlash).

It is worth noting that the PCM algorithm is applied by using each EMA sensor (resolver, RVDT, currents) as reference signal, and the analysis demonstrated that the technique accuracy increases when the motor angle is used (i.e., resolver).

The algorithm operates by recording, during a generic MBIT, the residue signal $\Delta\theta_m(t)$

$$\Delta\theta_m(t) = \left|\frac{\theta_{ms}(t)}{\tau_g} - \theta_{outd}(t)\right|, \qquad (4.56)$$

where θ_{ms} is the resolver output, τ_g is the overall gear ratio, and θ_{outd} is the output rotation demand.

The residue signal and the actuator demand are then post-processed by the following steps:

1. compute the FFT of both signals, i.e., $\Delta\Theta_m(\omega)$ and $\Theta_{outd}(\omega)$, where $\omega = 2\pi f$ and f indicates the frequency variable;
2. compute $\Delta\Theta_m(\omega)/\Theta_{outd}(\omega)$, i.e., normalize the FFT of the motor angle residue with respect to the demand, to valorize the onset of kinetic energy in frequency ranges where the position demand is not relevant;
3. multiply the normalized FFT data by the amplification factor

$$F_{amp}(\omega) = \left[\frac{|\theta_{ms}(j\omega)|}{\tau_g}|\theta_{outd}(j\omega)| - 1\right]^{-1}, \qquad (4.57)$$

   in which the ratio $|\theta_{ms}(j\omega)|/|\theta_{outd}(j\omega)|$ is obtained as the amplitude of the frequency response of motor angle with respect to actuator demand, to highlight the limit cycle content with respect to both high-frequency disturbances and low-frequency dynamics (Fig. 4.37);
4. the normalized and amplified FFT data are analyzed in the expected frequency range for the limit cycle onset (from 0.3 to 1 Hz), by searching for the peak of the distribution, Fig. 4.38;
5. the peak is assumed to identify the limit cycle characteristics, and its amplitude and frequency are calculated;
6. the measured amplitude and frequency are compared with the PCM database, Fig. 4.39a;

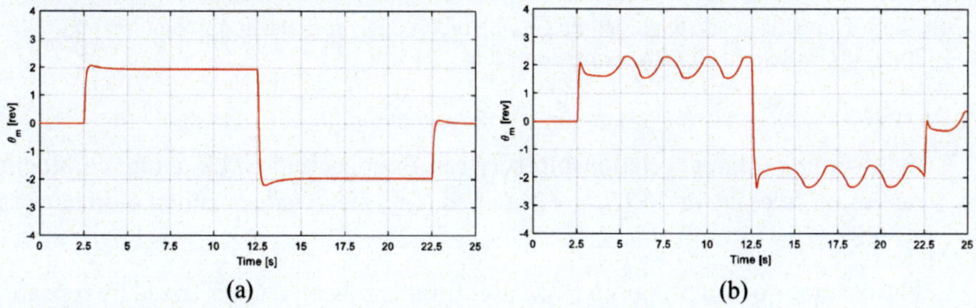

Fig. 4.36 Motor angle response during MBIT: **a** no freeplay; **b** end-life freeplay. © [2018] IEEE. Reprinted, with permission, from [17]

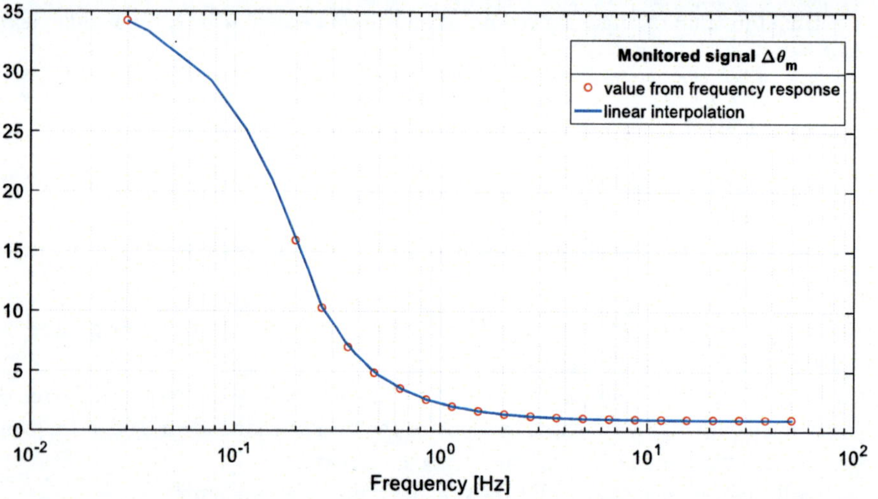

Fig. 4.37 Amplification factor of the FFT of the motor angle residue. © [2018] IEEE. Reprinted, with permission, from [17]

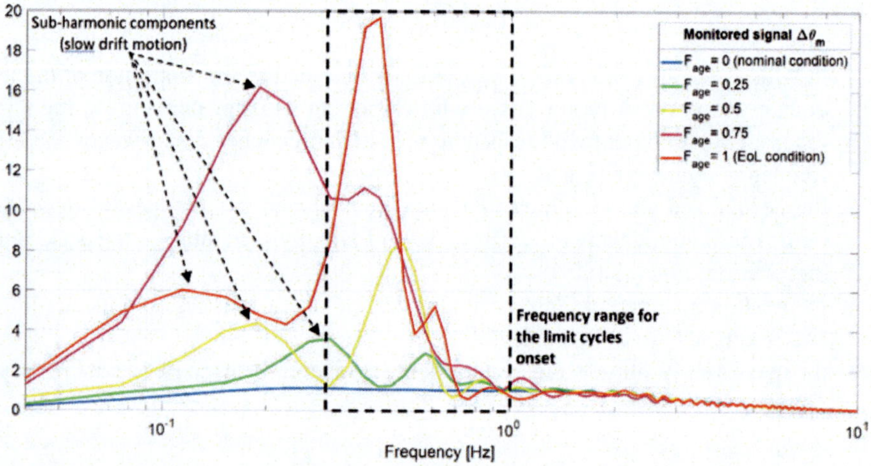

Fig. 4.38 Normalized and amplified FFT of the motor angle residual at different freeplay values. © [2018] IEEE. Reprinted, with permission, from [17]

7. by assuming a linear relationship between freeplay and EMA aging, confidence ranges on freeplay and RUL are obtained (i.e., mean values and maxima errors), 4.39b.

The accuracy of the proposed PCM algorithm has been characterized by repeating the MBIT simulation with different values of EMA freeplay, and by evaluating the effects of model uncertainties on the PCM outputs (thresholds in Fig. 4.39b).

The results reported in Fig. 4.40 demonstrate that the accuracy of the PCM algorithm is very good in the range of larger freeplays and old ages (from 70% age up to end-of-life, the maxima errors in terms of age are lower than 20% of the mean prog-

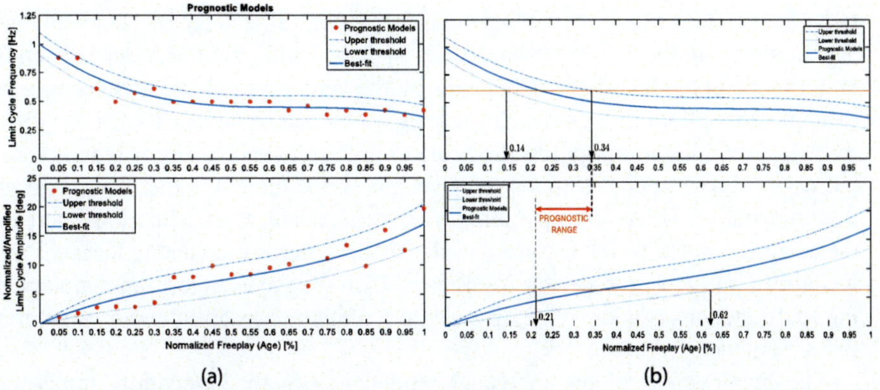

Fig. 4.39 **a** Database from prognostic models; **b** example of PCM results. © [2018] IEEE. Reprinted, with permission, from [17]

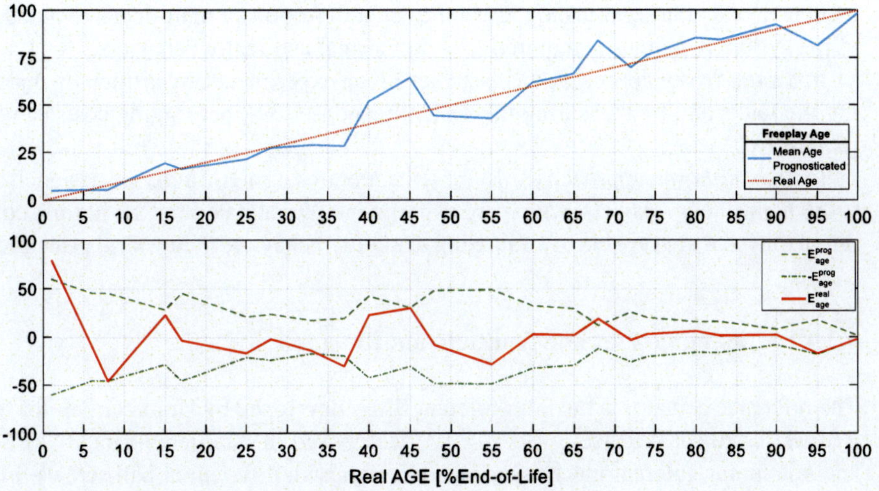

Fig. 4.40 Evaluation of the PCM accuracy. © [2018] IEEE. Reprinted, with permission, from [17]

nosticated value). On the other hand, the accuracy strongly decreases in the range of small or intermediate freeplays (lower than 50% age). This drawback is expected to be not critical, since the detrimental effects related to an intermediate freeplay are expected to be covered by the design safety factors.

4.3.3 Fault Diagnosis via High-Fidelity Dynamic Models

The survivability of jamming is one of the major challenges for researchers and engineers developing safety-critical aerospace EMAs, and the typical solution is to use redundant architectures, so that fail-operative and/or fail-safe systems are obtained. The number and the type of redundancies applied within the EMA depends on the target reliability, which in turn, depends on the aerospace vehicle architecture. For example, the split of flight movables into independent sub-surfaces, each one driven by a dedicated EMA, simplifies the actuator architecture in terms of redundancies, but it generally implies an increase of weight and volume. On the other hand, the development of EMAs with jamming-tolerant mechanical transmissions increases the design complexity, but it simplifies the actuator integration and reduces the total weight. In all cases, the design of efficient fault diagnosis algorithms implementing FDI functions is a key issue, in order to maintain operability and/or to revert to fail-safe mode.

The employment of an analytical redundancy method generally implies an increase in the development costs of an actuator. Apart from the need of limiting the number of additional sensors and algorithms, special attention must be paid to the design approach of the algorithm. Since nonlinearities, disturbances, environment and loads can significantly affect the actuator response, an in-depth knowledge of the system dynamics is required for both normal and faulty behaviors.

In a *data-driven* approach, this is achieved via experiments, by artificially injecting the faults and by measuring the system response [59], but rigging costs can be prohibitive.

In a *model-based* approach [3, 16, 65], the monitoring algorithms are designed by using high-fidelity models, which are experimentally-validated for the normal condition only and are capable of simulating the faulty behaviors by physical principles.

4.3.3.1 Jamming-Tolerant Transmission Kinematics

The reference actuator is the fault-tolerant EMA developed by UmbraGroup for the REPRISE project – Phase 2 (Sect. 4.2.3), composed of dual-redundant BLACMs and a jamming-tolerant mechanical transmission with differential ball screws, Fig. 4.15.

The kinematics is thus based on a speed-summing paradigm and the relationships between the motors' speed $\theta_1(t)$ and $\theta_2(t)$, the output speed $\dot{x}_o(t)$, and the screw shaft translational speed $x_{ss}(t)$ and rotational speeds $\dot{\theta}_{ss}(t)$ are given by Eqs. (4.58)–(4.59)

$$\begin{cases} \dot{\theta}_{ss}(t) = c_{11}\dot{\theta}_1(t) + c_{12}\dot{\theta}_2(t) \\ \dot{x}_{ss}(t) = c_{21}\dot{\theta}_1(t) + c_{22}\dot{\theta}_2(t), \end{cases} \tag{4.58}$$

$$\dot{x}_o(t) = b_1\dot{\theta}_1(t) + c_2\dot{\theta}_2(t), \tag{4.59}$$

where the kinematic coefficients c_{ij} (with $i = 1, 2$ and $j = 1, 2$) are functions of the pitches of the motors' screw-nut couplings p_{s1} and p_{s2}, while b_r (with $r = 1, 2$) also depend on the pitch of the output shaft screw p_{s3}

$$c_{11} = \frac{p_{s1}}{p_{s1} - p_{s2}}, \qquad c_{12} = -\frac{p_{s2}}{p_{s1} - p_{s2}}, \tag{4.60a}$$

$$c_{21} = \frac{p_{s1} p_{s2}}{2\pi\,(p_{s1} - p_{s2})}, \qquad c_{22} = -c_{21}, \tag{4.60b}$$

$$b_1 = \frac{p_{s1}\,(p_{s2} - p_{s3})}{2\pi\,(p_{s1} - p_{s2})}, \qquad b_2 = -\frac{p_{s2}\,(p_{s1} - p_{s3})}{2\pi\,(p_{s1} - p_{s2})}. \tag{4.60c}$$

4.3.3.2 Operation Modes and Fault-Tolerant Control

The EMA is position-controlled with three nested loops on current, motor speed, and output translation, and it is equipped with dual ECUs for the independent drive of the two motors. Each ECU is composed of two electronic boards: one dedicated to the closed-loop control and the other to the condition monitoring. To enable the measurement of relevant signals from both control and monitor boards, the sensors' system is composed of:

- n. 12 current sensors (n. 2 per each motor phase);
- n. 4 resolvers (n. 2 per motor);
- n. 2 LVDTs, to sense the output shaft translation;
- n. 2 cone-type proximity sensors, for the screw shaft translation, to be used by monitor boards only.

Theoretically, due to the speed-summing kinematics, there are infinite linear combinations of motors' motions generating the same output translation (4.59). For the system development, a selection of relevant EMA operation modes (i.e., combinations of motors' motions) has been made, see Table 4.16. In *active/stand-by* operations (ASB and SBA modes), one motor rotates and the other is held by the related brake, so that the screw shaft has a roto-translating motion (4.58). In *active/active* operations, both motors rotate and, depending on their speeds, the screwshaft motion can range from roto-translation to pure translation (AAPT mode) or pure rotation (AAPR mode), see (4.58).

Among all possible active/active modes, there exists one that implies a balanced power split among the motors in quasi-dynamic regime (i.e., constant speeds). This speeds' combination is given by

$$\dot{\theta}_1(t) = \alpha_{EP}\dot{\theta}_2(t), \tag{4.61}$$

where

$$\alpha_{EP} = -\frac{p_{s2}\,(p_{s1} - p_{s3})}{p_{s1}\,(p_{s1} + p_{s3})}. \tag{4.62}$$

The speed correlation provided by (4.61) is imposed by the actuator control laws in the normal operation mode of the system, which is defined as *Active-Active Equal Power* (AAEP) mode. Depending on the configuration of the two actuators, other modalities are possible, as reported in Table 4.16:

- AAPT: Active/Active Pure Translation;

Table 4.16 Operation modes of the fault tolerant-EMA

Motor 1 operation	Motor 2 operation	Screwshatf motion	Model acronym
Active	Active	Roto-translation	AAEP (Normal operation)
Active	Active	Pure translation	AAPT
Active	Active	Pure rotation	AAPR
Active	Braked	Roto-translation	ASB
Braked	Active	Roto-translation	SBA

- AAPR: Active/Active Pure Rotation;
- ASB: Active/Stand-By;
- SBA: Stand-By/Active.

The mechanical transmission thus permits to control the output position by tolerating one motor fault as well as one jamming to motors' ball-nuts or to the screw shaft, provided that the faults are correctly detected and isolated, and an appropriate switching of the operation mode is applied. It is actually worth noting that excessive FDI latencies can affect the whole aircraft safety, especially if the EMA is used for primary flight controls.

4.3.3.3 High-Fidelity Model Features

To design the fault diagnosis and condition monitoring algorithms, a detailed nonlinear model of the EMA dynamics is developed and validated with experiments in the normal operative condition (i.e., no faults). The model includes the simulation of:

- three-phase BLACMs, driven via FOC technique;
- SVPWM-driven MOSFET power bridge;
- nested digital loops on motors' currents, motors' speed, and output position;
- sensors' errors (bias, noise);
- control nonlinearities (saturation, rate limiting, quantization);
- 5-DoF mechanical transmission with equations of motions related to motors' rotation, output translation, and screw shaft rotation and translation;
- jamming faults generating the block of motors' rotation and screw shaft translation and rotation).

4.3.3.4 Model of the Mechanical Transmission with Dual Motors

A schematic representation of the 5-DoF mechanical transmission integrating the two EMA motors is reported in Fig. 4.41, and the resulting twelfth-order dynamic model of the fault-tolerant EMA is given by (4.63), in which the electrical equations are derived from a 1st-order equivalent monophase model of the BLACM (Sect. 4.3.2.3).

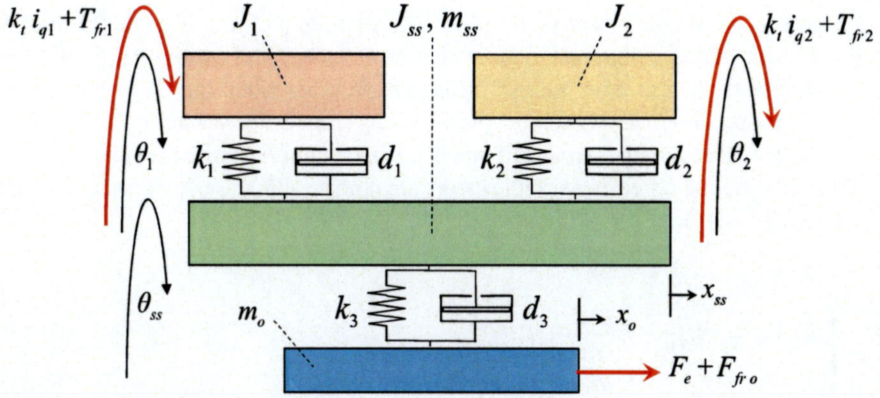

Fig. 4.41 Scheme of the 5-DoF mechanical transmission

Table 4.17 Parameters of the fault tolerant-EMA model

Parameter	Value	Unit
p_{s1}	-15×10^{-3}	m
p_{s2}	15×10^{-3}	m
p_{s3}	-3.175×10^{-3}	m
$J_1 = J_2$	6.5×10^{-3}	$kg \cdot m^2$
J_{ss}	0.39×10^{-3}	$kg \cdot m^2$
m_{ss}	0.9	kg
m_o	1.3	kg
R	0.41	V/A
L	2×10^{-3}	H
k_t	0.97	Nm/A
$k_1 = k_2$	3.67×10^5	Nm/rad
k_3	9.8×10^3	Nm/rad
$d_1 = d_2$	9.73	$Nm \cdot s/rad$
d_3	0.39	$Nm \cdot s/rad$

The quantities V_{q1} , V_{q2}, i_{q1}, and i_{q2} are the quadrant voltages and currents of the motors, J_1, J_2, and J_{ss} are the motors' and the screw shaft inertias, m_{ss} and m_o are the screw shaft and the output rod masses, R and L are the resistance and the inductance of the motor phases, k_t is the motors' torque constant, F_e is the external load, F_{fro} is the output rod friction, T_{fr1} and T_{fr2} are the motors' frictions, while k_m and d_m are the stiffness and damping of the m-th screw-nut coupling ($m = 1, 2, 3$). The parameters of the EMA model are reported in Table (4.17).

Excerpts of the model validation results are reported from Figs. 4.42, 4.43, 4.44 and 4.45 with reference to the AAEP mode only (similar results are obtained for the simulation of all EMA operation modes). The model errors during large-displacement

/ high-speed tests are lower than 1% of EMA stroke (0.5 mm, Fig. 4.42) and very good results are also obtained in terms of motors' speed and currents (Figs. 4.43 and 4.44), though the currents' prediction is less accurate when the EMA holds about 2 kN compression load at fixed position (results from 1.5 to 4 s in Fig. 4.44).

The model accuracy increases if small-displacement responses are considered (in sinusoidal frequency responses, the errors are comparable to sensors' accuracy up to 10 Hz, Fig. 4.45).

$$
\begin{cases}
V_{q1} = Ri_{q1} + L\dfrac{di_{q1}}{dt} + k_t\dot{\theta}_1 \\
V_{q2} = Ri_{q2} + L\dfrac{di_{q2}}{dt} + k_t\dot{\theta}_2 \\
J_1\ddot{\theta}_1 = k_t i_{q1} + T_{fr1} - d_1\left(\dot{\theta}_1 - \dot{\theta}_{ss} - \dfrac{2\pi}{p_{s1}}\dot{x}_{ss}\right) - k_1\left(\theta_1 - \theta_{ss} - \dfrac{2\pi}{p_{s1}}x_{ss}\right) \\
J_2\ddot{\theta}_2 = k_t i_{q2} + T_{fr2} - d_2\left(\dot{\theta}_2 - \dot{\theta}_{ss} - \dfrac{2\pi}{p_{s2}}\dot{x}_{ss}\right) - k_2\left(\theta_2 - \theta_{ss} - \dfrac{2\pi}{p_{s2}}x_{ss}\right) \\
J_{ss}\ddot{\theta}_{ss} = d_1\left(\dot{\theta}_1 - \dot{\theta}_{ss} - \dfrac{2\pi}{p_{s1}}\dot{x}_{ss}\right) + k_1\left(\theta_1 - \theta_{ss} - \dfrac{2\pi}{p_{s1}}x_{ss}\right) \\
\qquad + d_2\left(\dot{\theta}_2 - \dot{\theta}_{ss} - \dfrac{2\pi}{p_{s2}}\dot{x}_{ss}\right) + k_2\left(\theta_2 - \theta_{ss} - \dfrac{2\pi}{p_{s2}}x_{ss}\right) \\
\qquad - d_3\left[\dot{\theta}_{ss} + \dfrac{2\pi}{p_{s3}}(\dot{x}_{ss} - \dot{x}_o)\right] - k_3\left[\theta_{ss} + \dfrac{2\pi}{p_{s3}}(x_{ss} - x_o)\right] \\
m_{ss}\ddot{x}_{ss} = \dfrac{2\pi d_1}{p_{s1}}\left(\dot{\theta}_1 - \dot{\theta}_{ss} - \dfrac{2\pi}{p_{s1}}\dot{x}_{ss}\right) + \dfrac{2\pi k_1}{p_{s1}}\left(\theta_1 - \theta_{ss} - \dfrac{2\pi}{p_{s1}}x_{ss}\right) \\
\qquad + \dfrac{2\pi d_2}{p_{s2}}\left(\dot{\theta}_2 - \dot{\theta}_{ss} - \dfrac{2\pi}{p_{s2}}\dot{x}_{ss}\right) + \dfrac{2\pi k_2}{p_{s2}}\left(\theta_2 - \theta_{ss} - \dfrac{2\pi}{p_{s2}}x_{ss}\right) \\
\qquad - \dfrac{2\pi d_3}{p_{s3}}\left[\dot{\theta}_{ss} + \dfrac{2\pi}{p_{s3}}(\dot{x}_{ss} - \dot{x}_o)\right] - \dfrac{2\pi k_3}{p_{s3}}\left[\theta_{ss} + \dfrac{2\pi}{p_{s3}}(x_{ss} - x_o)\right] \\
m_o\ddot{x}_o = F_e + F_{fro} \\
\qquad + \dfrac{2\pi d_3}{p_{s3}}\left[\dot{\theta}_{ss} + \dfrac{2\pi}{p_{s3}}(\dot{x}_{ss} - \dot{x}_o)\right] + \dfrac{2\pi k_3}{p_{s3}}\left[\theta_{ss} + \dfrac{2\pi}{p_{s3}}(x_{ss} - x_o)\right]
\end{cases}
\tag{4.63}
$$

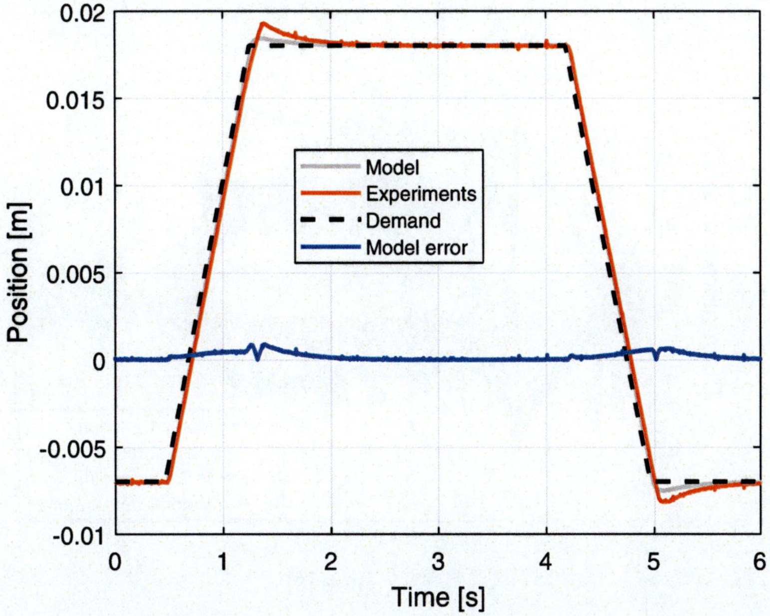

Fig. 4.42 Time-domain model validation: position tracking in AAEP mode with 100 N/mm spring-type loading (position response)

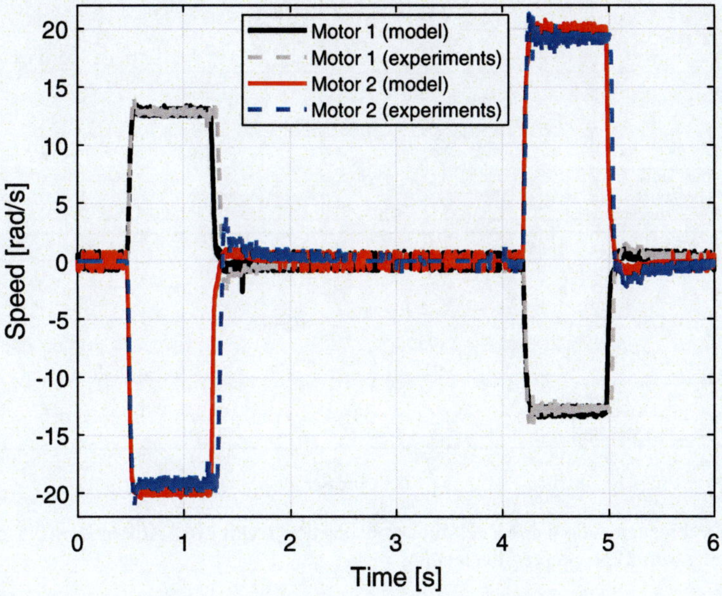

Fig. 4.43 Time-domain model validation: position tracking in AAEP mode with 100 N/mm spring-type loading (speed response)

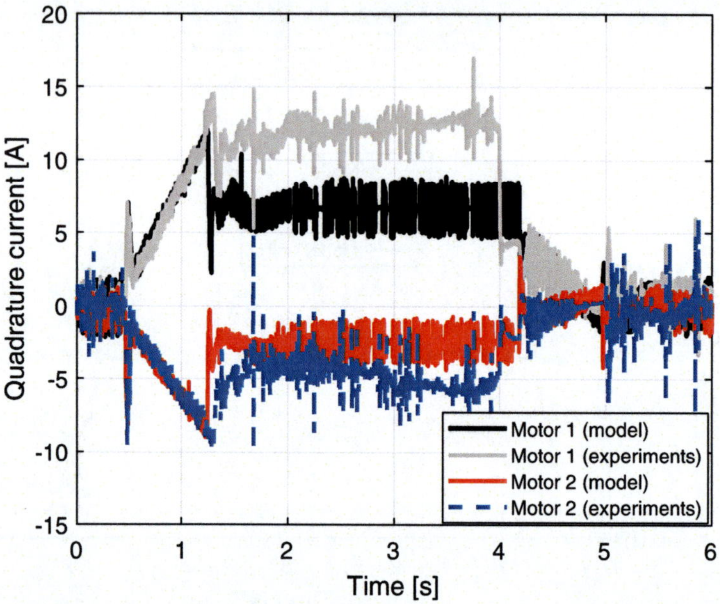

Fig. 4.44 Time-domain model validation: position tracking in AAEP mode with 100 N/mm spring-type loading (current response)

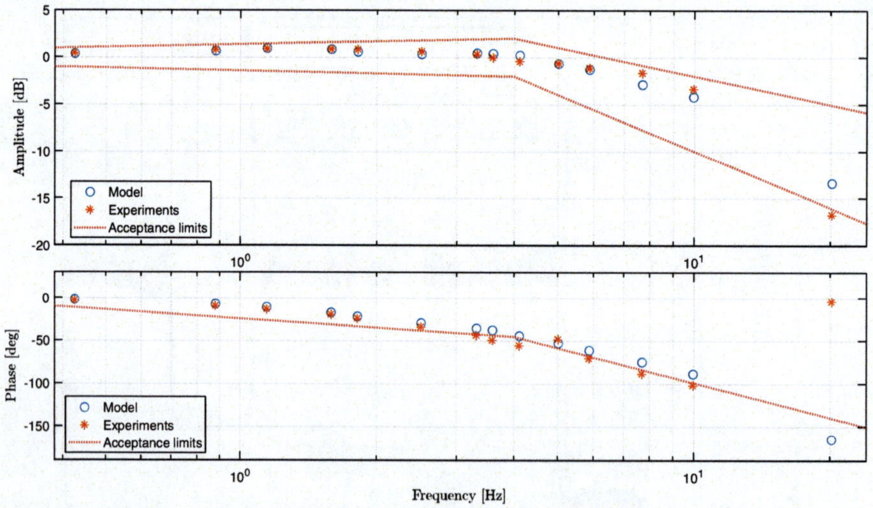

Fig. 4.45 Frequency-domain model validation: position tracking in AAEP mode at 0.5 mm demand amplitude, with 2 kN compression loading

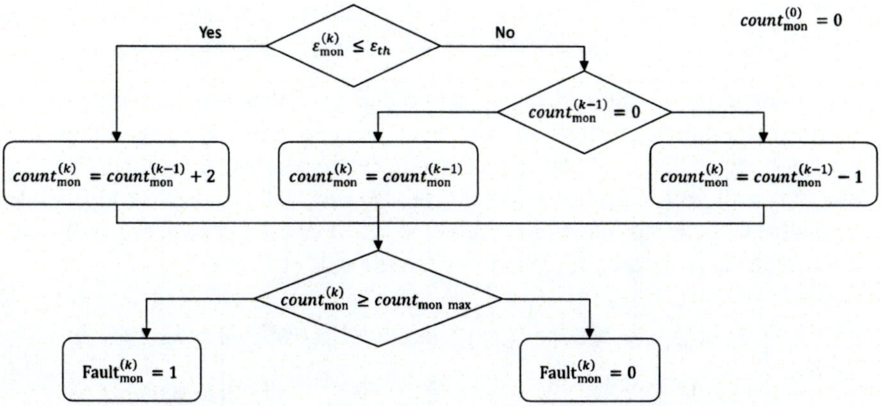

Fig. 4.46 Generalized jamming detection logic (JDL)

4.3.3.5 Jamming Monitoring Algorithms

The experimentally-validated model is thus used to design the fault diagnosis algorithms. With particular reference to the ones dedicated to the jamming FDI, they are composed of two sections executed in series: the first one targets the motors' jamming, and the second one focuses on the screw shaft jamming, which can result in a rotation or translation stuck. The jamming monitoring system aims to define, for each k-th monitoring sample ($k = 1, 2, \ldots$), the fault flag vector $\mathbf{Fault}_{\mathbf{mon}}^{(k)}$

$$\mathbf{Fault}_{\mathbf{mon}}^{(k)} = \left[\mathrm{Fault}_{\mathrm{mon}|1}^{(k)} \quad \mathrm{Fault}_{\mathrm{mon}|2}^{(k)} \quad \mathrm{Fault}_{\mathrm{mon}|ssRJ}^{(k)} \quad \mathrm{Fault}_{\mathrm{mon}|ssTJ}^{(k)} \right]^{\top} \tag{4.64}$$

where $\mathrm{Fault}_{\mathrm{mon}|1}$ and $\mathrm{Fault}_{\mathrm{mon}|2}$ are the fault flags related to motors' jamming, while $\mathrm{Fault}_{\mathrm{mon}|ssRJ}$ and $\mathrm{Fault}_{\mathrm{mon}|ssTJ}$ are the fault flags related to screw shaft rotation and translation jamming, respectively. If no jamming is detected, all the fault flag vector components are 0 (i.e., false Boolean values). Otherwise, if a jamming is detected at least one fault flag vector component is 1 (i.e., true Boolean value). The fault is isolated when only one fault flag is 1.

Each monitoring section uses a generalized Jamming Detection Logic (JDL), which is schematically reported in the flow chart of Fig. 4.46: the fault flag $\mathrm{Fault}_{\mathrm{mon}}$ is generated by elaborating a monitor signal $\varepsilon_{\mathrm{mon}}$ sampled at the monitoring frequency ($\mathrm{Fault}_{\mathrm{mon}}$). If the monitor signal is lower than a predefined threshold (ε_{th}, a fault counter ($count_{\mathrm{mon}}$) is increased by 2; if the threshold is exceeded, the fault counter is decreased by 1 if it is positive at the previous step, otherwise it is held at 0. The jamming is detected, when the fault counter exceeds a predefined value ($count_{\mathrm{mon}}$ max, which basically defines the detection latency).

In the algorithm related to the f-th motor jamming ($f = 1, 2$), the monitor signal $\varepsilon_{\mathrm{mon}|f}$ is defined as the variation of the motor rotation θ_f between two samples

$$\varepsilon_{\mathrm{mon}|f}^{(k)} = \left|\theta_f^{(k)} - \theta_f^{(k-1)}\right|, \tag{4.65}$$

and the algorithm operates according to (4.66): if the f-th motor rotation demand θ_{fd} between two samples varies for more than a predefined threshold ε_{df}, the jamming detection logic (JDL in (4.66)) is executed to define the related fault flag $\mathrm{Fault}_{\mathrm{mon}|f}$; otherwise, all fault flags are set to 0 (see also Fig. 4.47) and the AAEP mode is maintained. If a motor jamming is detected, the operation is switched from AAEP mode to the appropriate active/stand-by mode (ASB or SBA).

$$\begin{cases} \mathrm{Fault}_{\mathrm{mon}|f}^{(k)} = \mathrm{JDL}\left(\varepsilon_{\mathrm{mon}|f}^{(k)}\right) & \text{if} \quad \left|\theta_{fd}^{(k)} - \theta_{fd}^{(k-1)}\right| \geq \varepsilon_{df} \\ \mathbf{Fault}_{\mathbf{mon}}^{(k)} = \mathbf{0} & \text{if} \quad \left|\theta_{fd}^{(k)} - \theta_{fd}^{(k-1)}\right| < \varepsilon_{df} \end{cases} \tag{4.66}$$

In normal operative condition, the motors' speeds are correlated via (4.61), while, if there is a screw shaft rotation jamming ($\dot{\theta}_{ss} = 0$ in (4.58)), we have $\dot{\theta}_1 = \frac{p_{s2}}{p_{s1}}\dot{\theta}_2$. On the other hand, if the jamming causes the screw shaft translation stuck ($\dot{x}_{ss} = 0$ in (4.58)), we have $\dot{\theta}_1 = \dot{\theta}_2$. An estimation of the output speed, reconstructed from the motor resolvers signals via (4.59), can be effectively used for jamming monitoring, since we have

$$\dot{x}_{or} = b_1\dot{\theta}_1^{(k)} + b_2\dot{\theta}_2^{(k)} = \begin{cases} (b_1\alpha_{EP} + b_2)\,\dot{\theta}_2^{(k)} & \text{No jam} \\ \left(b_1\dfrac{p_{s2}}{p_{s1}} + b_2\right)\dot{\theta}_2^{(k)} & \text{if } \dot{\theta}_{ss} = 0 \\ (b_1 + b_2)\,\dot{\theta}_2^{(k)} & \text{if } \dot{x}_{ss} = 0. \end{cases} \tag{4.67}$$

From (4.59) applied to motors' speed demands and (4.67), an output speed residual can be thus obtained

$$\left|\dot{x}_{od}^{(k)} - \dot{x}_{or}^{(k)}\right| = \left|b_1\left(\dot{\theta}_{1d}^{(k)} - \dot{\theta}_1^{(k)}\right) + \left(\dot{\theta}_{2d}^{(k)} - \dot{\theta}_2^{(k)}\right)\right| \tag{4.68}$$

In normal conditions, this residual is small due to the speed tracking imposed by the EMA control laws, while it significantly increases when there is a screw shaft jamming. Actually, by substituting (4.67) into (4.68), in case of screw shaft rotation jamming ($\dot{\theta}_{ss} = 0$), the speed residual is

$$\left|\dot{x}_{od}^{(k)} - \dot{x}_{or}^{(k)}\right| = \left|b_1\left(\alpha_{Ep} - \frac{p_{s2}}{p_{s1}}\right)\dot{\theta}_2^{(k)} + (b_1\alpha_{EP} + b_2)\left(\dot{\theta}_{2d}^{(k)} - \dot{\theta}_2^{(k)}\right)\right|, \tag{4.69}$$

while, in case of screw shaft translation jamming ($\dot{x}_{ss} = 0$), we have

$$\left|\dot{x}_{od}^{(k)} - \dot{x}_{or}^{(k)}\right| = \left|b_1\left(\alpha_{Ep} - 1\right)\dot{\theta}_2^{(k)} + (b_1\alpha_{EP} + b_2)\left(\dot{\theta}_{2d}^{(k)} - \dot{\theta}_2^{(k)}\right)\right|. \tag{4.70}$$

In both (4.69)–(4.70), the second contributions at second hands are minor (due to motors' speed control tracking), so that the speed residual can be approximated by (4.71)

$$\left|\dot{x}_{od}^{(k)} - \dot{x}_{or}^{(k)}\right| \approx \begin{cases} 0 & \text{No jam} \\ \left|b_1 \left(\alpha_{Ep} - \dfrac{p_{s2}}{p_{s1}}\right) \dot{\theta}_2^{(k)}\right| & \text{if } \dot{\theta}_{ss} = 0 \\ \left|b_1 \left(\alpha_{EP} - 1\right) \dot{\theta}_2^{(k)}\right| & \text{if } \dot{x}_{ss} = 0, \end{cases} \tag{4.71}$$

which emphasizes that any screw shaft jamming implies that the speed residual is large. The *fault isolation* (i.e., rotation or translation stuck) is finally obtained by executing two algorithms in parallel, based on the monitor signals in (4.72))

$$\varepsilon_{\text{mon}|ssRJ}^{(k)} = \left|\dot{\theta}_1^{(k)} - p_{s2}/p_{s1}\dot{\theta}_2^{(k)}\right|, \tag{4.72a}$$

$$\varepsilon_{\text{mon}|ssTJ}^{(k)} = \left|\dot{\theta}_1^{(k)} - \dot{\theta}_2^{(k)}\right|. \tag{4.72b}$$

In conclusion, the screw shaft jamming FDI (second section of the monitoring system) is performed by applying (4.73)

$$\mathbf{Fault}_{\mathbf{mon}}^{(k)} = \begin{cases} \mathbf{0} & \text{if } \left|\dot{x}_{od}^{(k)} - \dot{x}_{or}^{(k)}\right| < \varepsilon_s \\ \begin{bmatrix} 0 \\ 0 \\ JDL\left(\varepsilon_{\text{mon}|ssRJ}^{(k)}\right) \\ JDL\left(\varepsilon_{\text{mon}|ssTJ}^{(k)}\right) \end{bmatrix} & \text{if } \left|\dot{x}_{od}^{(k)} - \dot{x}_{or}^{(k)}\right| \geq \varepsilon_s. \end{cases} \tag{4.73}$$

The flow chart representing the complete jamming monitoring system is reported in Fig. 4.47, while the parameters of the algorithms are given in Table 4.18.

4.3.3.6 Failure Transients Characterization

The experimentally-validated model of the fault-tolerant EMA is finally used to simulate the jamming faults and to characterize the effectiveness of the condition monitoring system for the FDI and the fault compensation via operation mode switching. Excerpts of results are reported in Figs. 4.48, 4.49, 4.50, 4.51, 4.52, 4.53, 4.54, 4.55, 4.56, 4.57, 4.58, and 4.59, in which the EMA responses in terms of position, motors' speed, and motors' currents are plotted with and without the jamming compensation. All the simulations are carried out by commanding the EMA, under 2.4 kN compression load, to track a large-displacement demand (±18 mm) at maximum speed, and by injecting a jamming fault when the actuator reaches its midstroke position (i.e.,

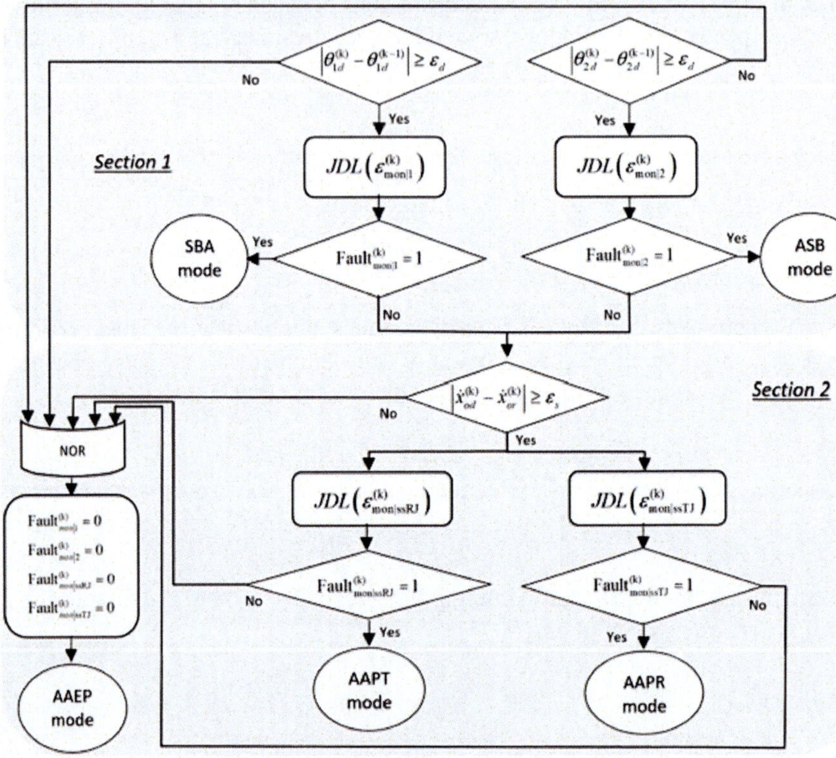

Fig. 4.47 Jamming monitoring system

at 2.03 s). The responses in case of jamming of the motor 1, causing the operation to switch from AAEP to SBA mode, are reported in Figs. 4.48, 4.49 and 4.50. The very small fault latency (8 ms), together with reconfigured demand speed for the active motor quite similar to the one before the fault (Fig. 4.49), generates a negligible failure transient. It is worth noting that, for this case, the jamming is not problematic for the output position tracking (see Fig. 4.48 without compensation), but the high current values in the blocked motor (see Fig. 4.50 without compensation) would rapidly cause an uncontrolled heating with possible extension of the fault consequences to control boards and/or the active motor. Similar considerations can be made for the case of jamming to the motor 2, causing the operation to switch from AAEP to ASB mode (Figs. 4.51, 4.52 and 4.53), but in this case, the control laws reconfiguration allows to strongly reduce the position tracking error with respect to the uncompensated condition (Fig. 4.51). The failure transient related to the screw shaft translation jamming, causing the operation to switch from AAEP to AAPR mode, is then shown in Figs. 4.54, 4.55 and 4.56. In this case, the jamming effects without compensation are dramatically negative, since the EMA reacts by diverging in opposite direction to the position demand up to its endstroke (Fig. 4.54). In addition, high currents are

Table 4.18 Parameters of the jamming monitor algorithms

Parameter	Value	Unit
f_{mon}	2000	Hz
$\varepsilon_1 = \varepsilon_2$	2.5×10^{-3}	rad
$\varepsilon_{th\|1} = \varepsilon_{th\|2}$	0.05	rad
$count_{\text{mon max}\|1} = count_{\text{mon max}\|2}$	16	–
ε_s	0.12×10^{-3}	m/s
$\varepsilon_{th\|ssRJ}$	2	rad/s
$\varepsilon_{th\|ssTJ}$	1	rad/s
$count_{\text{mon max}\|ssRJ} = count_{\text{mon max}\|ssTJ}$	20	–

generated in both motors (Fig. 4.56) and consequent heating issues arise. On the contrary, the jamming monitoring algorithms, intervening with very small latency (12 ms), allow to maintain the EMA under control with small tracking error. Finally, the results related to the screw shaft rotation jamming, causing the operation to switch from AAEP to AAPT mode, are given from Figs. 4.57, 4.58 and 4.59. In this case, the fault latency is more concrete (82 ms), but no variation of tracking performance is observable (Fig. 4.57). Similarly to previous fault cases, the jamming does not cause the loss of the position tracking (see Fig. 4.57 without compensation), but it is dangerous for the motors, due to the high current values (see Fig. 4.59 without compensation).

The results thus demonstrate that the model-based jamming monitoring system, composed of two sections in series dedicated to motors and differential screws respectively, succeeds in detecting and isolating faults at different locations of the mechanical transmission, with minor failure transients. The monitoring is designed by using a high-fidelity nonlinear model of the system derived from physical first principles (model accuracy is from 0.2 to 1% of the EMA full-stroke, depending on input amplitudes and operational frequency, see).

4.3.4 Final Considerations on Model-Based Approaches

Model-based approaches represent the most structured approach to fault diagnosis. By including a specific model and relations for the input (which can encompass both known commands, disturbances, and faults) and output signals in the considered system, it is possible to have proven performance about the detection and isolation of the faults. The specific fault modeling assumptions permit the use of related techniques, for which properties have been largely studied in the literature. Therefore, if a discrete-quality model of the system can be obtained, model-based approaches are the most robust solution to employ, since they permit to decouple (or considerably attenuate) the effect of external disturbances on the residual signal. Thus, they can

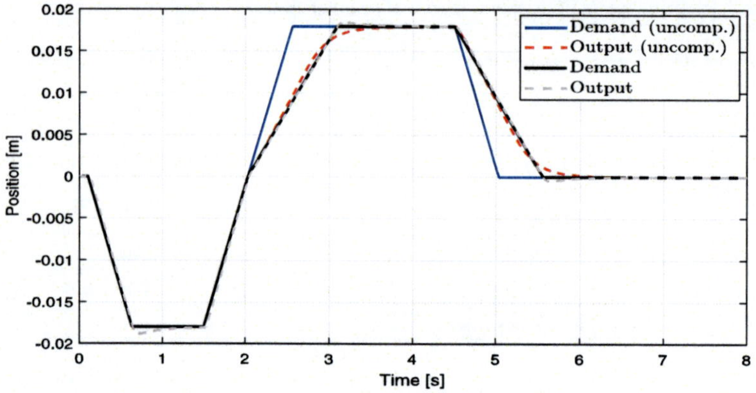

Fig. 4.48 Motor 1 jamming at 2.03 s: position responses (fault latency = 8 ms)

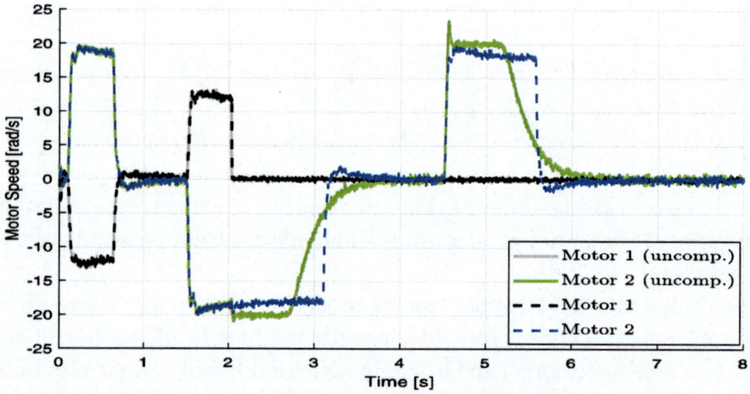

Fig. 4.49 Motor 1 jamming at 2.03 s: speed responses (fault latency = 8 ms)

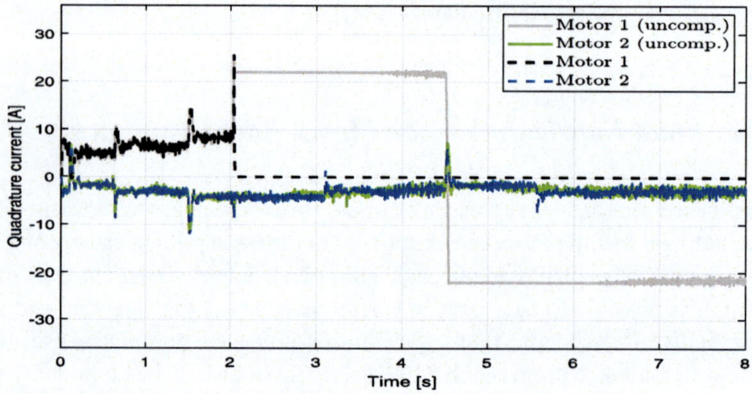

Fig. 4.50 Motor 1 jamming at 2.03 s: current responses (fault latency = 8 ms)

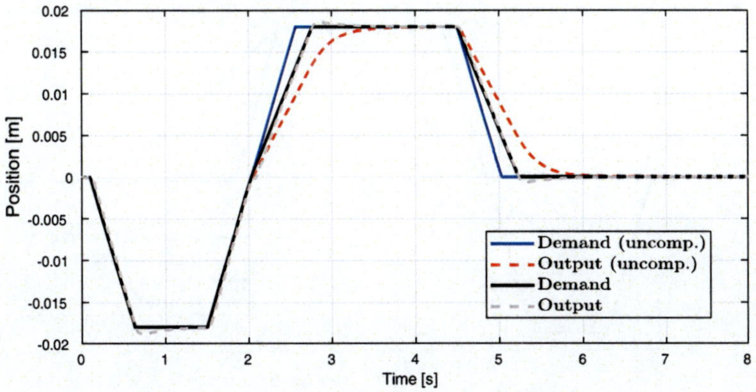

Fig. 4.51 Motor 2 jamming at 2.03 s: position responses (fault latency = 8 ms)

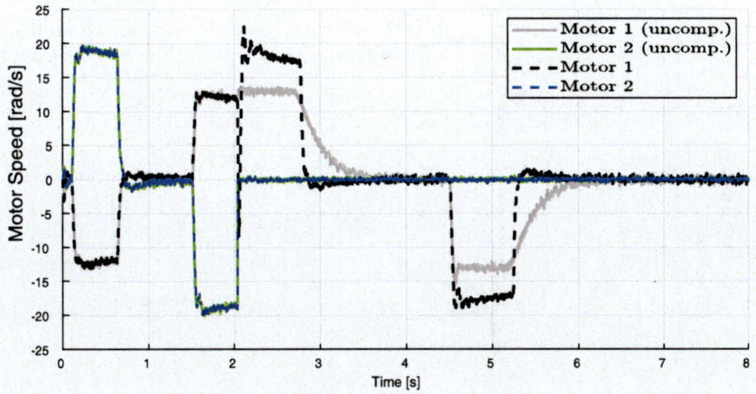

Fig. 4.52 Motor 2 jamming at 2.03 s: speed responses (fault latency = 8 ms)

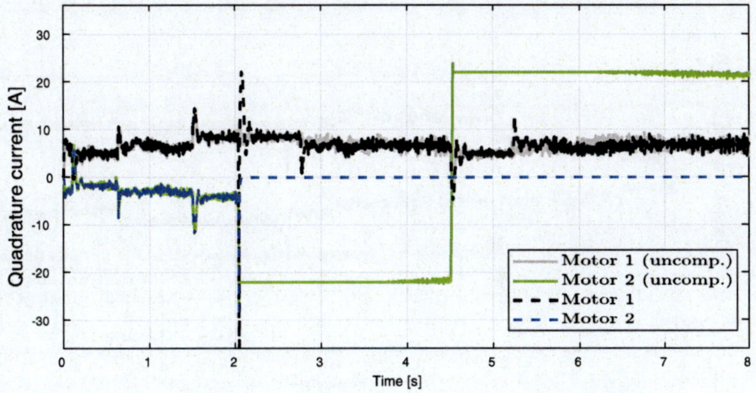

Fig. 4.53 Motor 2 jamming at 2.03 s: current responses (fault latency = 8 ms)

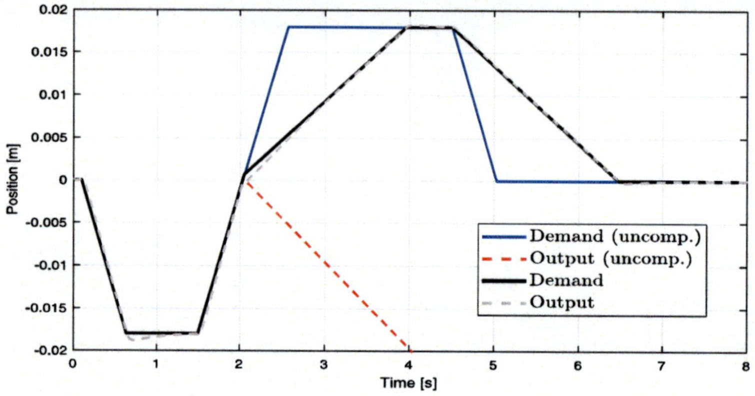

Fig. 4.54 Screw shaft translation jamming at 2.03 s: position responses (fault latency = 8 ms)

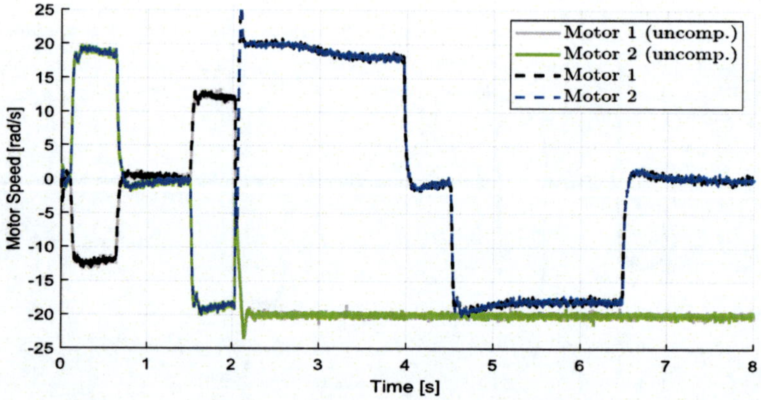

Fig. 4.55 Screw shaft translation jamming at 2.03 s: speed responses (fault latency = 8 ms)

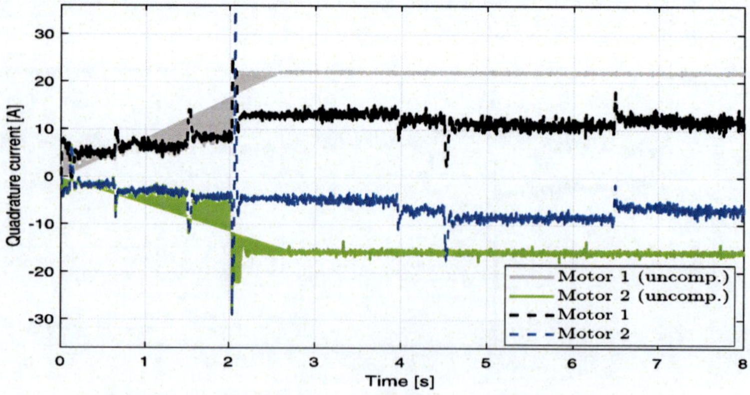

Fig. 4.56 Screw shaft translation jamming at 2.03 s: current responses (fault latency = 8 ms)

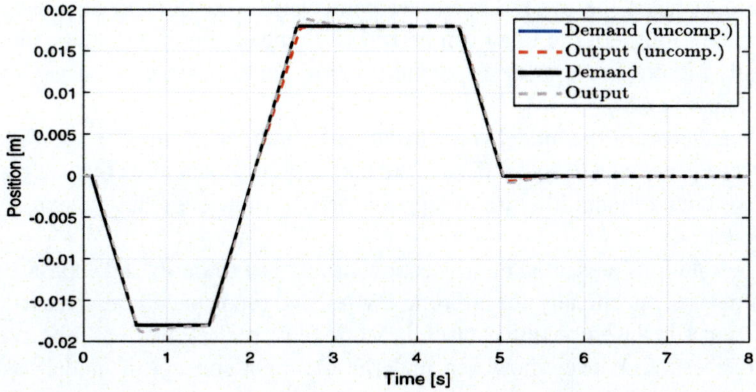

Fig. 4.57 Screw shaft rotation jamming 2.03 s: position responses (fault latency = 8 ms)

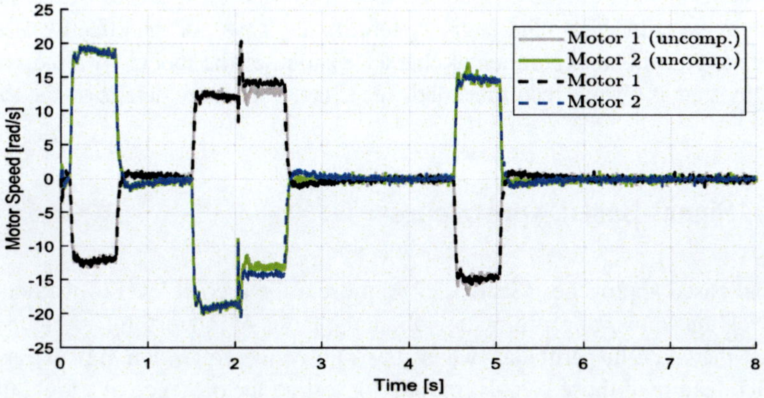

Fig. 4.58 Screw shaft rotation jamming 2.03 s: speed responses (fault latency = 8 ms)

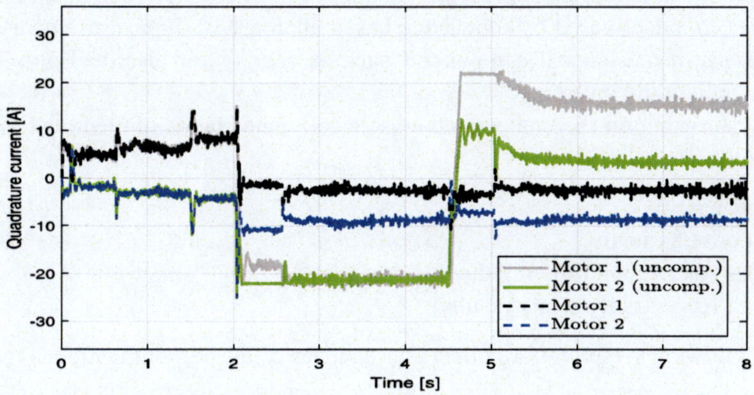

Fig. 4.59 Screw shaft rotation jamming 2.03 s: current responses (fault latency = 8 ms)

work also when the system operating point varies (contrary to knowledge-based approach). Furthermore, they can seamlessly manage multiple inputs and outputs signals, without looking for a specific symptom in each signal (contrary to signal-based approaches).

Nonetheless, if the model is not accurate enough, or requires excessive time and cost resources to be developed, or if the fault exhibits symptoms that are difficult to capture with a model, the use of signal-based or knowledge-based approaches can be preferred.

The solutions presented in this section, devoted to model-based approaches, have been devised specifically for airborne EMAs, and represent alternatives to standard approaches in the automatic control literature (e.g., parity equations, observer-based approaches, etc.). They, however, underlie the main concept of model-based fault diagnosis approaches, i.e., the generation of residuals sensitive to faults.

As outlined in this section, the use of a system model can be seen not only as a method for fault diagnosis, but also as a substitute for a dedicated test bench to perform experiments on the faulty system. In any case, for the effectiveness of the whole approach, an experimental characterization of the model with the system in healthy state is always recommended (see Chap. 5 for more details on the topic).

4.4 Signal-Based Approaches

Signal-based approaches (Sect. 3.2.2) represent the second level of methodologies that rely on some prior information about the faults of interest, following to model-based methods. In particular, when it is known *where* (which signals) and *what* (which feature of those signals) to look for testing the presence of a fault, then it is possible to employ these methods.

In general, given the specificity of the prior knowledge required about faults symptoms, signal-based approaches are applied at a narrower component level with respect to model-based or knowledge-based approaches. Typical examples are the diagnosis of mechanical components such as bearings or electrical ones such as transformers and inverters.

By focusing on the components-wise level, major faults of electrical machines mainly occurs at [46]:

- bearings;
- stator or armature;
- rotor bar and end ring of induction machines;
- shaft (eccentricity-related faults).

The following symptoms are direct conseguence of the before mentioned fault:

- unbalanced air gap voltages and line currents;
- increased torque pulsations;
- decreased average torque;

- increased losses and reduction in efficiency;
- excessive heating.

Specific signals can be employed to detect those symptoms, by employing techniques such as:

- electromagnetic field monitoring;
- temperature measurements;
- infrared recognition;
- radio frequency (RF) emissions monitoring;
- vibration monitoring;
- acoustic noise measurements;
- motor current signature analysis (MCSA).

In particular, the analysis of the currents in electrical motors has received much attention in the literature [46, 57]. We now review the main types of faults that can be diagnosed by employing signal-based methods.

4.4.1 Common Faults in Electro-Mechanical Actuators Diagnosable by Signal-Based Approaches

4.4.1.1 Bearing Faults

Rolling (or balls) bearings are mechanical components whose function is to interpose between machine parts in mutual rotation and to limit their friction. They are one of the most widely used elements in machines and their failure one of the most frequent reasons for machine breakdown. Rolling elements can have different geometries: spheres, needles, cylindrical, tapered or barrel rollers. The choice of the type of rolling element depends on the application, the load to which it is subjected, and its direction.

They are composed of several elements:

- *Races:* the surfaces on which the bearing rolls. The load placed on the bearing is supported by this contact surface. In general, the inner ring rests on the shaft, while the outer ring rests on the bearing housing;
- *Rolling elements:* the rolling elements are constructed in such a way as to allow also their rotation (simultaneous rotation around their axis and around the axis of the bearing);
- *Cage:* separates the rolling elements at a regular interval, holds them in position between the internal and external races, and allows their free rotation.

Types of bearing defects. There exist a vast categorization of the possible faults that can affect the bearing functionality. We provide here a list of the most notable ones:

- *Galling*: type of wear due to the friction that occurs when moving two materials, that are compressed on each other;

- *Spalling*: process by which metal is broken into small fragments (spalls).
- *Peeling*: formation of coarse irregularities on the surface of the coating applied to a metal surface;
- *Pitting*: corrosion that presents itself as localized attacks, by creating small holes on the surface of the metal;
- *Scoring*: surface damage caused by debris accumulated in the bearing under improper lubrication conditions or excessive loads;
- *Smearing*: superficial damage that occurs due to the presence of small debris between the bearing components due to the breakage of the lubricant film or due to slippage of the elements;
- *Fracture and cracks*: fracture of the elements can be caused by excessive or impulsive load acting locally on the component considered;
- *Denting*: it occurs when the debris, made up of small metal particles, are in the contact area between the rotating element and the track;
- *Fretting*: wear that occurs due to repeated slipping between two surfaces. Fretting occurs both on the mounting surfaces and on the contact surfaces between the tracks and the rotating elements;
- *Creep*: is a phenomenon in which slipping between two mounting surfaces creates a play;
- *Seizure*: when the bearing overheats quickly during rotation, the bearing changes color. After overheating the tracks, the rotating elements and the cage slowly begin to melt and deform, accumulating more and more damage;
- *True brinelling*: occurs when the load on the bearing is greater than the elastic limit of the bearing material;
- *False brinelling*: it looks similar to true brinelling but is due to vibrations. For example, during transport, vibrations can cause the rolling elements to move, and therefore, leave indentations on the tracks;
- *Flaking*: occurs when small particles of material detach from the surface of the rolling track or from the rolling body due to fatigue, forming rough and irregular areas.

Diagnosis of bearing faults. The causes of bearing defects can be of a different nature: wear (e.g., due to lack of lubricant/ maintenance), fatigue, excessive loads, presence of debris, incorrect installation, and misalignments. We can classify bearing defects into two major classes:

- *Localized defects*: defects localized in a specific position or element of the bearing (e.g., cracks, incisions);
- *Distributed defects*: defects that span multiple elements of a wide portion of a single element (e.g., misalignments, eccentricity of the races or rolling elements).

Localized defects are often indicators of failures in progress. Their monitoring is, therefore, crucial.

Specific *fault frequencies* which amplitude is enhanced by the defect can be analytically computed from the geometry of the bearing and it revolution speed. These are the *symtptoms* that we can look for in an accelerometer signal, placed orthogonal to the rotational axes of the bearing. The vibration frequencies (in Hz) to detect these faults are given by

$$Outer\,race\,defect: \qquad f_v = \frac{n_b}{2} f_r[1 - b_d \cos(\beta)/d_p] \tag{4.74a}$$

$$Inner\,race\,defect: \qquad f_v = \frac{n_b}{2} f_r[1 + b_d \cos(\beta)/d_p] \tag{4.74b}$$

$$Ball\,defect: \qquad f_v = \frac{d_b}{2b_d} f_r[1 - b_d \cos(\beta)/d_p]^2 \tag{4.74c}$$

$$Train\,(cage)\,defect: \qquad f_v = \frac{1}{2} f_r[1 - b_d \cos(\beta)/d_p] \tag{4.74d}$$

where f_r is the rotational frequency, n_b is the number of balls, b_d and d_p are the ball diameter and ball pitch diameter, respectively, and β is the contact angle of the ball (with the races). In the case of *induction motors*, these vibration frequencies reflect themselves in the current spectrum as

$$f_{bng} = |f_{supply} \pm m \cdot f_v|, \tag{4.75}$$

where f_e is the electrical supply frequency [57].

It is important to notice that the analysis of fault frequencies (4.74) is meaningful when the rotational speed is *constant*. If this is not the case, *order analysis* can be performed [6]. In any case, those symptoms are best found by looking in vibration signals acquired by accelerometers (usually piezoelectric ones), and processed by means on an *envelope analysis* followed by a spectrum of the demodulated signal [52].

4.4.1.2 Screw and Nut Assembly

The most feared event for screw/nut assembly is the jamming, that is produced usually by a severe heating of both components during the operation [26]. This can be caused by:

- lubrication failure;
- recirculating jam of rotating elements;
- degradation of races/rolling elements surfaces due to wear.

The lubrication and recirculation jam failures appear as a sudden increment of the friction torque, while the fatigue failure appears as a gradual increment of the backlash. The friction torque increments can be measured using the current values; the backlash can be diagnosed by comparing the input position with the output position (taking into account the effect of the dynamic system given by the motor).

4.4.1.3 Stator or Armature Faults

These faults are usually related to insulation failure (also known as phase-to-ground or phase-to-phase faults). The majority of stator windings fail as a result of gradual deterioration of the electrical insulation, caused, e.g., by partial discharges (PD) phenomena. PD are small voltage sparks that occur in high voltage insulation wherever small air pockets exist. The monitoring of PD can be performed online by means of specific measurements circuits and sensors [60].

Another approach, specific for low-voltage induction machines, is the use of an external flux sensor to detect interturn short circuits [25]. In this case, it is possible to look for specific frequencies, related to the turn-to-turn fault, in the dispersion (axial) flux measurements

$$f_s = (k \pm n(1-s)/p) f_{supply}. \tag{4.76}$$

where p is the number of pole pairs, f_{supply} is the main supply frequency, $k = 1, 3$, $n = 1, 2, \ldots (2p-1)$ and and s is the slip. This detection technique has been shown to be more reliable than MCSA, especially when the number of shorted turns is small compared to the total number of turns in a phase winding.

In a Brushless DC (BLDC) motor, stator (windings) faults result from the breakdown of the winding isolation [32], which occurs from the overheating of the coil due to Joule effect. Monitoring of winding temperature can be effective in detecting the fault (usually estimated from the phase currents).

4.4.1.4 Broken Rotor Bar Faults

Broken rotor bar in induction motors has been faced in [20] by using MCSA. They investigated the sideband components around the fundamental supply frequency

$$f_b = (1 \pm 2ks) f_{supply}, \quad k = 1, 2, 3, \ldots; \tag{4.77}$$

while the lower sidebands are specifically due to a broken bar, the upper sidebands are due to consequent speed oscillation. Axial flux and torque measurements can also be employed.

4.4.1.5 Eccentricity-Related Faults

Machine eccentricity is the condition of unequal air gap that exists between the stator and rotor. We can distinguish between static and dynamic eccentricity. In case of static eccentricity, the position of the minimal radial air gap length is fixed in space. In case of dynamic eccentricity, the center of the rotor is not at the center of the rotation and the position of minimum air gap rotates with the rotor. The presence of static and dynamic eccentricity can be detected using MCSA [1, 52]. The equation describing the frequency components of interest is

$$f_e = f\left[(kR \pm n_d)\frac{(1-s)}{p} \pm \nu\right], \tag{4.78}$$

where the *eccentricity order* is $n_d = 0$ in case of static eccentricity, and $n_d = 1, 2, 3, \ldots$, in case of dynamic eccentricity, f_{supply} is the fundamental supply frequency, R is the number of rotor slots, s is the slip, p is the number of pole pairs, k is any integer, and ν is the order of the stator harmonics that are present in the power supply driving the motor ($\nu = \pm 1, \pm 3, \pm 5, \ldots$).

Vibration signals can also be monitored to detect eccentricity-related faults in induction motors [7]. In case of mixed eccentricity, the low-frequency stator vibration components are given by

$$f_v = 2f_{supply} \pm f_r. \tag{4.79}$$

4.4.1.6 Electronics

The electronic subsystem includes power converters and switching bridges for motor operation (BLDC motors). The monitoring of the voltage drop on voltage stabilizing capacitors and the on-resistance of switching transistors for winding supply control is an important aspect of electronics components diagnosis [26]. A somewhat "hardware" approach can also be suggested, such as the application of down-sized transistors which operate in similar conditions as for the main transistors. The breakdown of this transistor is a precursor event for the damage in the main one.

4.4.2 *Example: Fault Detection and Isolation of Bearing Defects*

We now present a practical application of diagnosing an inner race pitting fault in a ball bearing used in workcenter machines. The aim is to evaluate the vibration signal measured from a piezoelectric accelerometer placed on the housing of the bearing, in an orthogonal direction with respect to the bearing rotation axis. So, the accelerometer directly measures the ball's passages. The sampling frequency for the accelerometer was set to 12800 Hz.

4.4.2.1 Symptoms of Localized Faults

When a rolling element (such as the balls) "go across" a (locally) damaged element, it is like it gets "hit" or "excited" by an impulsive input, just like when a car steps over a speed bump on the road. If we could measure a vibration, we will see those hits

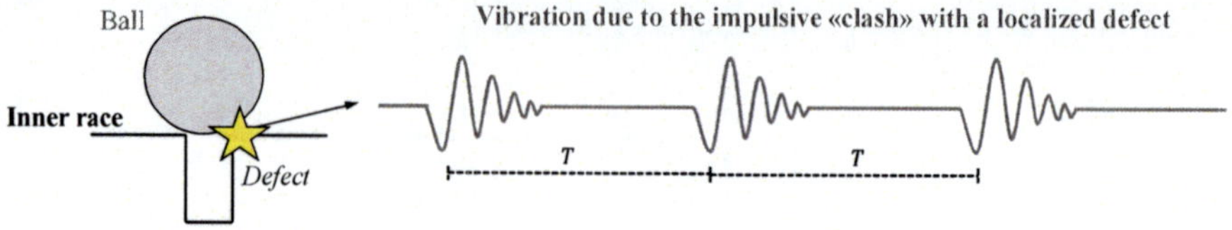

Fig. 4.60 Inner race bearing fault. The rolling element hits the faulty area every T time instants (related to the fault frequency) and produces a modulation of the bearing vibration behavior

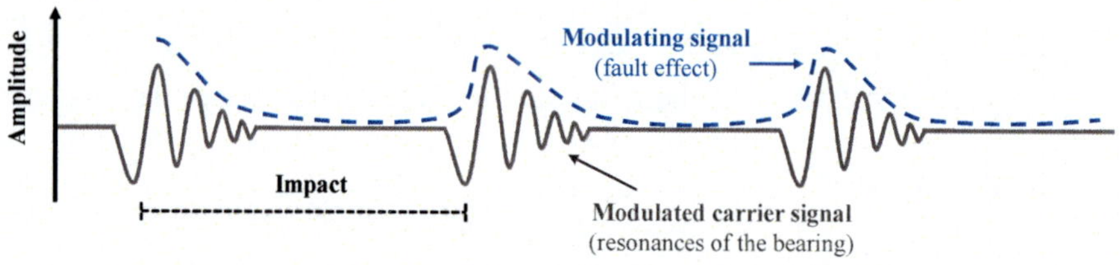

Fig. 4.61 Diagnosing a bearing race fault can be viewed as an amplitude demodulation problem

in correspondence with the faulty surface. Local faults in rolling element bearings produce a series of impacts which repeat (almost) periodically at a rate dependent on bearing geometry and rotation speed, see (4.74). The fault *modulates* (almost) periodically the standard operating vibrations of the bearing.

The diagnostic *information of interest is contained in the repetition frequency* of the impact series (Fig. 4.60), and not in the frequency spectrum resulting from the impacts, as this would usually be a sum of the excited bearing resonance frequencies. In fact, vibration signals are often severely corrupted by strong levels of background noise, encompassing all other vibration sources in the machine under inspection. The recognition of these impulses is made more difficult if we consider that also random fluctuations in the shaft speed compromise the repeatability of the fault impulses responses.

This problem can be formulated as that of detecting transient signals in strong additive noise. Impactive faults excite the structural resonances of the bearing, simply amplifying standard operational vibrations. This variation effect on the amplitude of the natural frequency is known as *amplitude modulation*, Fig. 4.61. The computation of the signal envelope allows us to be more robust compared to variations in the "distance" between the pulses (due to changes in contact angle and loads).

The aim of *envelope analysis* (also known as amplitude demodulation) is to reconstruct the modulating signal from the measured modulated one. Then, a frequency analysis of the resulting modulating signal can be performed to evaluate the presence of the fault frequency (and its harmonics).

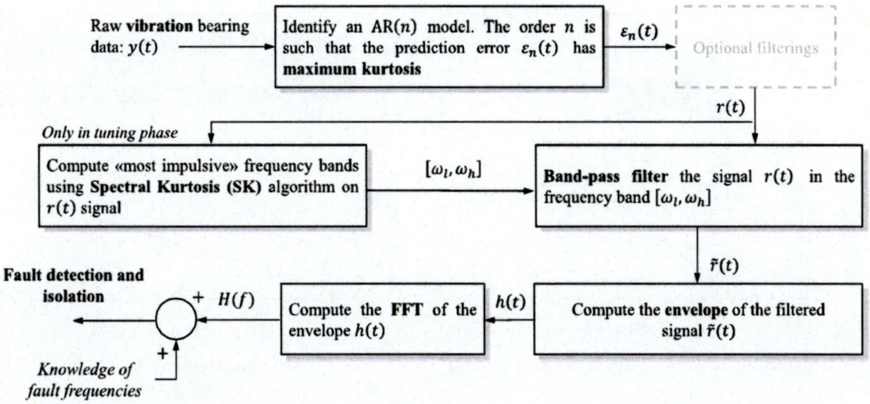

Fig. 4.62 Bearing diagnosis steps

4.4.2.2 A Bearing Diagnosis Flowchart

The typical steps for bearing diagnosis are presented in Fig. 4.62. The design of the algorithm is as follows:

1. Firstly, constant rotation speed tests as to be performed, in order to not mix the measurements with vibrations due to multiple speeds and frequencies. If it is not possible to perform constant speed data, order tracking techniques can be used (this will require also the measurement of the rotation speed) [51].
2. Starting from constant speed accelerometer measures $y(t)$, the raw signal goes through a series of filtering steps. The first one consists of identifying and AutoRegressive (AR) or order n, such that the prediction error $\varepsilon_n(t)$ of the model has maximum kurtosis. Then the residual can be further filtered by additional (and more optional) procedures such as Adaptive Noise Cancellation (ANC), Time Synchronous Averaging (TSA) or Minimum Entropy Deconvolution (MED), see [52]. We call the resulting signal $r(t)$.
3. Only during the design phase (i.e., with data from a faulty bearing), the Spectral Kurtosis (SK) algorithm is used to automatically find the frequency bands where most of the "impulsive events" lie. This bandwith $[\omega_l, \omega_h]$ will then be used to filter the signal $r(t)$, obtaining the signal $\tilde{r}(t)$.
4. Then, the envelope $h(t)$ of the signal $\tilde{r}(t)$ is computed, using standard procedures such as the Hilbert-Huang Transform (HHT).
5. As the last step, the Fast Fourier Transform (FFT) $H(f)$ of the envelope $h(t)$ is computed, to look for specific fault frequencies, given the prior knowledge about the symptoms.

STEP 1: Obtain constant speed data. The first step consists of performing a test at constant rotational speed (or that includes a constant speed trajectory). In the example, we performed a test on a bearing mounted at the head of a ball screw

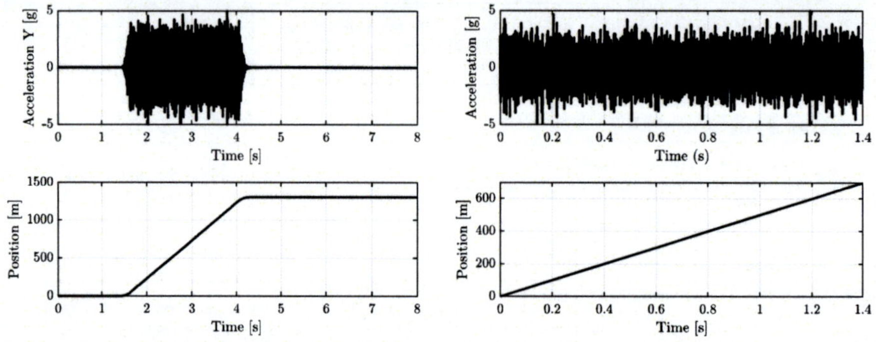

(a) Raw accelerometer and encoder data. (b) Constant speed data.

Fig. 4.63 Extraction of constant speed vibration data from a faulty bearing

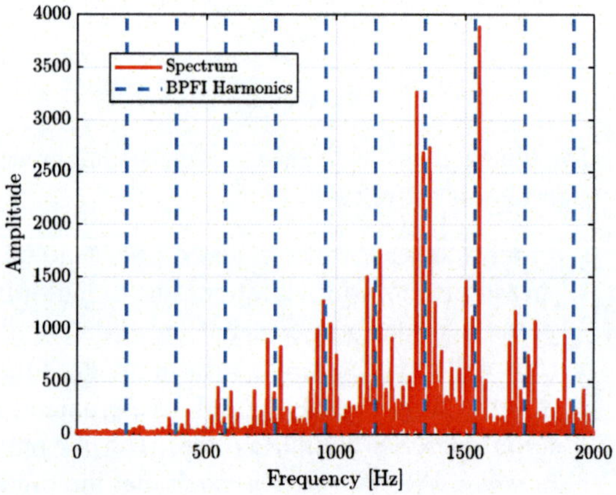

Fig. 4.64 Spectrum of the raw vibration signal, with overlapped the BPFI frequency and its harmonics

transmission, with a stroke at 30 m/min. Given the transmission ratio of 20 mm/round, this corresponds to about 25 Hz of rotation frequency.

Figure 4.63 shows the raw vibration data measured from a faulty inner race bearing, along with the position trajectory performed by the machine shaft. Only the portion of data that corresponds to a movement with constant speed is retained for further processing: this avoids a mixture of frequency components, due to the faut, to be present in the data.

For the current bearing, the BPFI (the fault frequency corresponding to an inner race fault), is equal to 192 Hz. In Fig. 4.64, it is possible to notice how performing a FFT on the raw signal $y(t)$ does not lead to clearly find the BPFI frequency.

STEP 2: Autoregressive filtering. Autoregressive (AR) models are typically used in the context of time-series modeling and econometrics. An AR(n) model, where n is the model order, is of the form

$$y(t) = a_1 y(t-1) + a_2 y(t-2) + \ldots + a_n y(t-n) + e(t) = A(z)e(t), \quad (4.80)$$

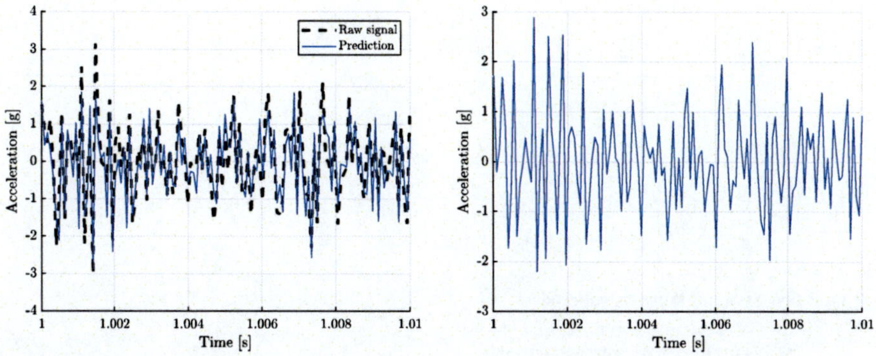

(a) One-step prediction $\hat{y}(t|t-1)$ of the AR(n^*) model.

(b) Residual ε_{n^*} of the AR(n^*) model.

Fig. 4.65 Autoregressive linear prediction and residual

where $a_1, \ldots, a_n \in \mathbb{R}$ and $e(t) \sim \text{WN}(0, \lambda^2)$ is a white noise signal with zero mean and variance λ^2. The term $A(z)$ is a polynomial in the lag-variable z (such that $zy(t) = y(t-1)$) and reads as

$$A(z) = 1 - a_1 z^{-1} - a_2 z^{-2} - \ldots - a_n z^{-n}. \tag{4.81}$$

The optimal one-step prediction $\hat{y}(t|t-1)$ is given by [35]:

$$\hat{y}(t|t-1) = [1 - A(z)]\, y(t), \tag{4.82}$$

from which a *prediction error* $\varepsilon_n(t)$ can be computed

$$\varepsilon_n(t) = y(t) - \hat{y}(t|t-1). \tag{4.83}$$

The aim of fitting an AR model to the constant speed vibration data $y(t)$ is to try to predict and remove the predictable component of the signal. The unpredictable component (the residual signal $\varepsilon_n(t)$) will then contain mostly "unpredictable events", such as the impulses due to the fault. Thus, linear prediction with an AR model is a way to enhance the fault-to-noise ratio.

In order to choose the order n of the AR model, [52] suggests to try a grid of orders (e.g., from 1 to 50), and then select the order n^* which gives the *highest kurtosis*, that represents a measure of "peakedness" of the signal.

The residual $\varepsilon_{n^*}(t)$ can then be further filtered and enhanced with other techniques [52]. The output of this step is the signal $r(t)$ (which can also be equal to $\varepsilon_{n^*}(t)$). Results of this step, with $n^* = 5$, can be found in Fig. 4.65.

STEP 3: Spectral kurtosis filtering. Spectral kurtosis (SK) provides a means of determining which frequency bands contain a signal of maximum impulsivity. These

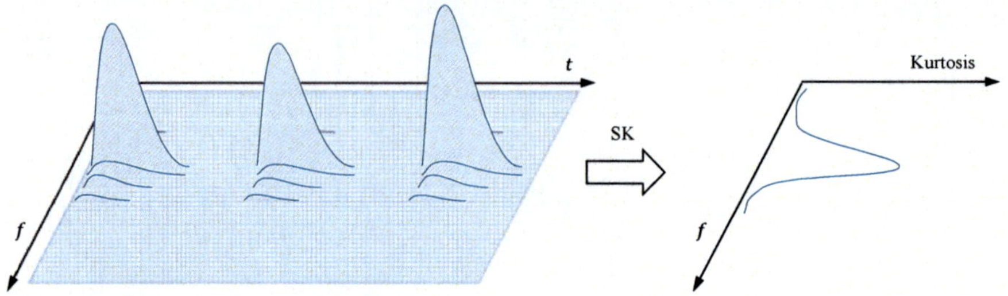

Fig. 4.66 Spectral kurtosis scheme

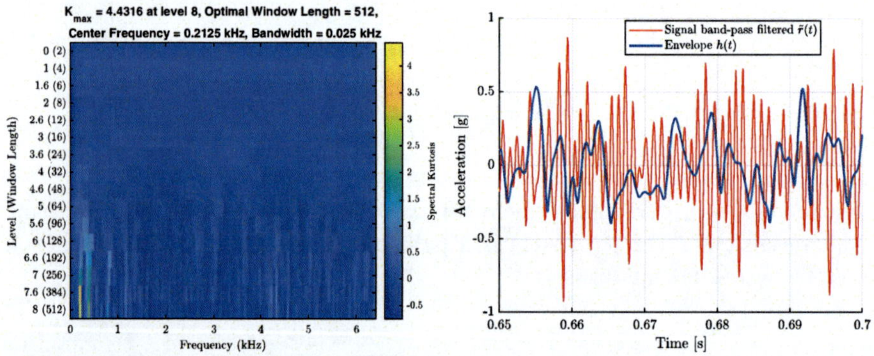

Fig. 4.67 Application of spectral kurtosis and envelope computation

impulsive behaviors (that give origin to transients such as those outlined in Fig. 4.61) should be attributed to the bearing faults.

Instead of computing the kurtosis of the whole time-domain signal, the SK algorithm divides the spectrogram of the signal (i.e., a time-frequency representation of how the frequency content varies over time) into "frequency bands". For each of these frequency bands, the kurtosis with respect to time is computed. The result is a kurtosis as function of the frequency, see Fig. 4.66.

From here, the frequency band $[\omega_l, \omega_h]$ with the highest kurtosis can be extracted, or designed. Figure 4.67-(a) presents the results of the MATLAB command `kurtogram`, applied to the signal $r(t)$. The method computes the kurtosis for different frequencies and frequency resolutions (length of the frequency window to consider for the computation), then automatically selects a center frequency and a frequency band for later filtering. These indications should be taken as indication and slightly adapted to the specific data. Figure 4.67-(b) reports the signal $\tilde{r}(t)$, i.e., $r(t)$ bandpass filtered in the bandwith $[\omega_l, \omega_h]$.

STEP 4: Envelope computation. As shown in Fig. 4.64, the spectrum of the raw signal often contains little diagnostic information about bearing faults. A consolidated technique is that of envelope analysis, where a signal is first bandpass filtered in a high frequency band, (in which the fault impulses are amplified by structural resonances), and then it is amplitude demodulated to form the envelope signal, whose

Fig. 4.68 Spectrum of the envelope signal, with overlapped the BPFI frequency and its harmonics

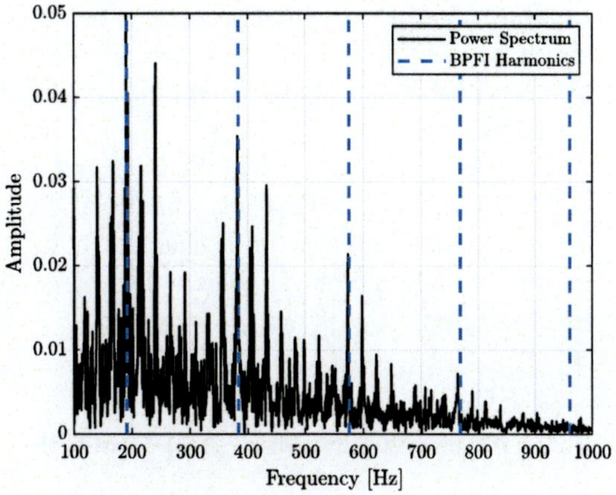

spectrum contains the desired diagnostic information in terms of both repetition frequency (ball pass frequency or ball spin frequency), as well as modulation by the appropriate frequency at which the fault is passing through the load zone (or moving with respect to the measurement point).

The envelope computation can be performed by several algorithms, such as the Hilbert-Huang Transform (HHT) [51]. Figure 4.67-(b) report the computed envelope $h(t)$ along with the signal $\tilde{r}(t)$ filtered in the bandwith devised with the SK algorithm.

STEP 5: Frequency analysis of the envelope. The final step consists of taking the FFT $H(f)$ of the envelope $h(t)$ and plotting its magnitude spectrum. Figure 4.68 shows how the fault frequency is clearly visible at the BPFI frequency and its multiples. Furthermore, the modulation sidebands due to the rotation of the shaft can be observed.

Once the spectrum of the envelope has been computed, several rationales can be used for raising an alarm. First, one has to choose a threshold, based on the amplitude of the faults frequencies in the faulty and healthy cases. Then, for detection, the first several harmonics can be considered, and if they all exceed the threshold, we can raise an alarm. An alternative solution is to raise an alarm when only at least one (but not all) of the harmonics are above the defined threshold.

4.4.3 Final Considerations on Signal-Based Approaches

Signal-based approaches are a viable method every time that:

1. there is knowledge about the effect that a fault produces in a determined set of signals or physical quantities;
2. those signals are easily measurable by some sensors.

In all of these cases, signal-processing techniques can be leveraged to enhance the signature of the fault inside the signal. The prior knowledge about the fault symptoms usually permits not only the detection, but also the isolation and the identification of the faults.

Concerning EMAs, most of the work in the literature has been done in the context of MCSA for induction motors, while very few results are present for BLDC motors. The most viable solution for mechanical faults is the use of accelerometers, although relying on only motor variables will achieve lower costs and space saving on the EMA envelope.

4.5 Knowledge-Based Approaches

This section gathers three knowledge-based approaches employed in the context of the HOLMES (Sect. 4.2.1) and REPRISE (Sect. 4.2.2) projects.

Contrary to model-based and signal-based approaches, knowledge-based methods do not have at disposal any prior information about the faults, see Sect. 3.2.3. In the model-based case, the prior information is given by the system model, and it is effectively leveraged by the model-based design methods such as the parity space or diagnostic observer design schemes [18]. Signal-based approaches rely on the knowledge about the symptoms: they know what to look for and also where to search (in what signals) for detecting the fault presence. Knowledge-based approaches [19] only assume that the faults leave "traces" of their presence in the data. Both the what and where have to be learned or discovered by using a high amount of historical data about the system, in healthy and (possibly) on faulty conditions.

In particular, we present:

- a fault detection and isolation approach for ball screw balls fault, based on machine learning classifiers in Sect. 4.5.1 (HOLMES project);
- a condition monitoring approach for ball screw transmissions based on change-point detection methods in Sect. 4.5.2 (REPRISE project - phase 1);
- a condition monitoring approach for ball screw transmissions based on statistical process monitoring techniques in Sect. 4.5.3 (REPRISE project - phase 1).

4.5.1 Knowledge-Based Fault Detection and Isolation via Machine Learning Techniques

This section presents a knowledge-based approach for tackling the fault detection problem of the HOLMES project, introduced in Sect. 4.2.1.

Casting a fault diagnosis problem as a (supervised) machine learning one basically consists of computing indicators from raw measurements (e.g., the features of the problem) and then perform a classification of the features into "classes". Those classes may define the system's health and faulty states. The classification algorithm is first trained on experimental data, for which the true system state is known (class labels). Then, the classifier is evaluated on another set of data not used for training.

In what follows, several classification algorithms are evaluated on features computed from only actuator data, especially the phase current measurements [24].

4.5.1.1 Supervised Machine Learning Fault Detection Strategy

The classification algorithm takes as input a set of features, computed from a batch of data, obtained by a sliding window approach. A natural problem is the selection of the window length. Usually, a sensitivity analysis is performed, by evaluating a trade-off between the computational time and the number of measurements needed to obtain a reliable estimate of the feature values. In this case, a value of 1.5 s, with an overlapping factor of 0.75 s has been chosen, but these hyperparameters have to be tuned for the application at hand.

The following features are computed for each window of data, following the works of [5, 10, 33, 53, 66]:

1. torque to load ratio;
2. root mean square value;
3. kurtosis;
4. skewness;
5. frequency power in (0–50 Hz];
6. peak-to-valley;
7. energy operator;
8. crest factor;
9. shape factor;
10. mean frequency;
11. frequency center;
12. root Mean Square frequency;
13. standard deviation frequency;
14. sixth central moment;
15. mean value of the EMA surface temperature.

All features (apart from the first and the last) are computed using the motor quadrature current. The computation of the first feature consists of the ratio of the motor torque with respect to the load cell measurement. In practical applications, the features 1 and 15 will not be available: however, it is interesting to study their effect on the classification decision.

The output is a feature matrix $\mathbf{X} \in \mathbb{R}^{N \times m}$, where $N = 5359$ is the number of observations and $m = 15$ is the number of features.

Table 4.19 Classifiers comparison summary. Reproduced from [38] by permission. © IFAC 2017

Classifier	Mean Test F1-score	Mean Nested CV F1-score	Std. error Nested CV F1-score
Logistic regression	0.25	0.21	0.024
Support Vector Machines	0.70	0.70	0.005
Naive Bayes	0.13	0.12	0.006
Gradient Tree Boosting	0.83	**0.82**	0.009

4.5.1.2 Design and Evaluation of the Machine Learning Classifier

Once the feature matrix is formed, it is possible to split the data into train (80%) and test (20%) set. This lead to a total of 4287 training data and 1072 test data. As a first stage, a robust standardization procedure is applied to the train data matrix [54]. This standardization is applied also to test data by employing the same quantities used to standardize the training data.

Several classifiers were evaluated. The selection of the classifiers' hyperparameters has been performed via fivefold cross-validation using the training data. The best hyperparameters are then selected to train the respective models.

Table 4.19 reports the performance of the selected classifiers, evaluated on the test dataset and with a nested cross-validation procedure on the entire dataset [8]. The weighted mean F1-score is reported [49] (where the weights are given by the prevalence of a certain class in the considered dataset). The nested CV approach also allows to obtain an estimate of the standard deviation of the mean F1-score over different data folds. An higher value means that the discovered hyperparameters are not reliable, and so is the model.

The best classifier is the Gradient Tree Boosting (GTB) algorithm, which outperforms simpler models. Furthermore, the performance result seems to be stable as denoted by the standard deviation. Figure 4.69 represents the features that contribute more to the decisions of the GTB model. The most informative ones are the cage temperature, the torque to load ratio, and the computed frequency content.

Clearly, the first two most important features for the classification of the faults will rarely be available on a flight EMA; netherveless, this work suggests that the EMA quadrature current could be used for fault diagnosis purposes.

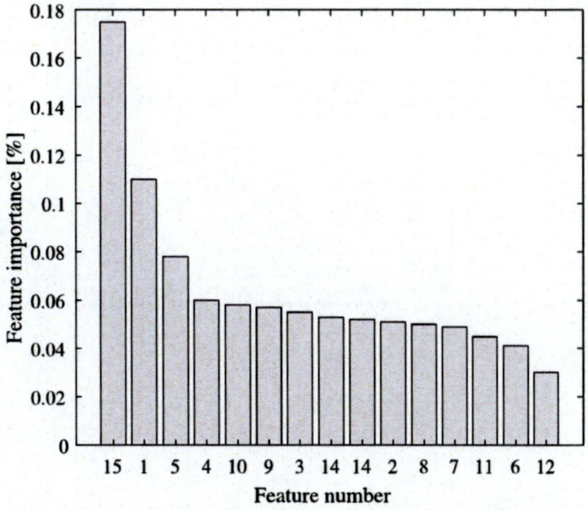

Fig. 4.69 Features importance. Reproduced from [38] by permission. © IFAC 2017

4.5.2 Knowledge-Based Condition Monitoring via Change Detection Algorithms

This section is devoted to a condition monitoring algorithm that relies on change-point detection techniques. The aim of those methods is to assess, at every time stamp, if the distribution of the data has statistically changed with respect to the previous time stamp. Naturally designed to be online, change-point detection methods can be thought as providing a continuous assessment of the properties of the data distribution.

The proposed approach characterizes itself as a *preflight strategy*, i.e., a test that can be performed before aircraft takeoff. This ensures that external disturbances (such as aerodynamic load) are kept as controlled as possible. In order to achieve this, it is possible to transform a change-point detection method to work in a *batch* mode, that can be used to compare two different experiments in time. Detected changes in the data distribution are symptoms of changes in the system behavior, and can be effectively used as health state indicators. The approach is tested on the REPRISE phase 1 EMA (Sect. 4.2.2.2).

4.5.2.1 Change Detection for Online Data

Among the many different approaches devised for change detection, one of the most appreciated ones is the one based on the Relative unconstrained Least-Squares Importance Fitting (RuLSIF) method [29, 34]. The RuLSIF approach directly estimates the *ratio* of two data densities [61].

In order to use the RuLSIF method for change detection, the data distributions before and after a certain time instants are taken as the two main quantities over which to estimate their ratio. Suitable divergences can be used to assess their dissimilarity

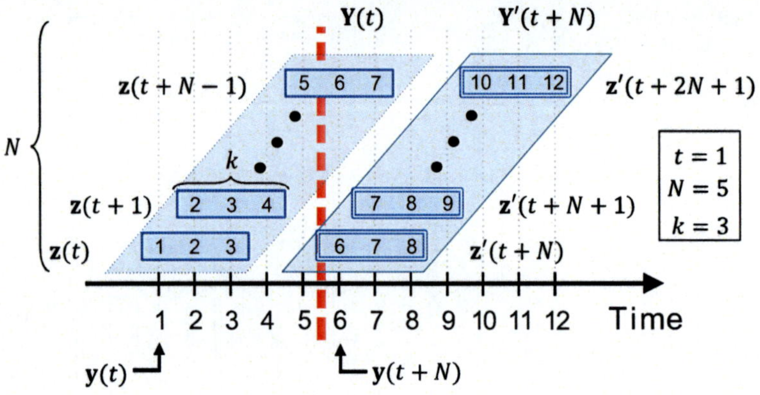

Fig. 4.70 Online RuLSIF data management example, with $m = 1, t = 1, N = 5$ and $k = 3$. © IEEE 2018. Reprinted, with permission, from [41]

[22]. A bounded (relative) density ratio estimator was introduced in [34] in order to avoid errors in the density ratio estimate (e.g., undefined values).

Denote $\mathbf{y}(t) \in \mathbb{R}^{m\times 1}$ as a m-dimensional time-series sample at time t, and let

$$\mathbf{z}(t) \equiv \left[\mathbf{y}(t)^\top,\ \mathbf{y}(t+1)^\top,\ \cdots,\ \mathbf{y}(t+k-1)^\top\right]^\top \in \mathbb{R}^{m\cdot k\times 1}.$$

The quantity $\mathbf{z}(t)$ will be treated as a single observation. The quantity $\mathbf{Y}(t)$ forms an Hankel matrix, composed by N observations of the mk-th dimensional samples $\mathbf{z}(t)$, starting from t:

$$\mathbf{Y}(t) \equiv [\mathbf{z}(t),\ \mathbf{z}(t+1),\ \cdots,\ \mathbf{z}(t+N-1)]^\top \in \mathbb{R}^{N\times m\cdot k}.$$

and it is widely employed in change-point detection methods that rely on subspace learning [30].

Let $\mathbf{Y}(t)$, $\mathbf{Y}(t+N)$ be two consecutive data segments. A change is detected if a dissimilarity measure between $\mathbf{Y}(t)$, $\mathbf{Y}(t+N)$ crosses a certain threshold. Figure 4.70 summarizes the main quantities of the method.

Pearson divergence for dissimilarity computation. Let now:

- P and P' the probability distributions of the data in $\mathbf{Y}(t)$ and $\mathbf{Y}(t+N)$, respectively;
- $p(\mathbf{x})$, $p'(\mathbf{x})$ the probability density functions of P and P', respectively.

The Pearson divergence is defined as [48]:

$$\mathrm{PE}\left(P \parallel P'\right) \equiv \frac{1}{2}\int p'(\mathbf{x}) \cdot \left(\frac{p(\mathbf{x})}{p'(\mathbf{x})} - 1\right)^2 d\mathbf{x}, \tag{4.84}$$

where $\mathbf{x} \in \mathbb{R}^{mk\times 1}$ represents a generic random variable.

Let now $\{\mathbf{z}_i\}_{i=1}^N$ and $\{\mathbf{z}'_j\}_{j=1}^N$ be a set of N samples drawn from $p(\mathbf{x})$ and $p'(\mathbf{x})$, respectively. The divergences (4.84), is computed by using *estimate* of the density ratio $\frac{p(\mathbf{x})}{p'(\mathbf{x})}$, from samples which are representative of the two distributions. The samples $\{\mathbf{z}_i\}_{i=1}^N$ are those belonging to $\mathbf{Y}(t)$. The samples $\{\mathbf{z}'_j\}_{j=1}^N$ are those belonging to $\mathbf{Y}(t+N)$.

The $\tilde{\alpha}$-relative Pearson divergence measure, introduced in [34] due to its robustness to numerical problems, is defined for $0 \le \tilde{\alpha} < 1$ as:

$$\begin{aligned} \mathrm{P}E_{\tilde{\alpha}}\left(P \,\|\, P'\right) &\equiv \mathrm{PE}\left(P \,\|\, \tilde{\alpha} P + (1-\tilde{\alpha})\, P'\right) \\ &= \frac{1}{2}\int p'_{\tilde{\alpha}}(\mathbf{x}) \cdot \left(\frac{p(\mathbf{x})}{p'_{\tilde{\alpha}}(\mathbf{x})} - 1\right)^2 d\mathbf{x}, \end{aligned} \tag{4.85}$$

where $p'_{\tilde{\alpha}}(\mathbf{x}) = \tilde{\alpha}\, p(\mathbf{x}) + (1-\tilde{\alpha})\, p'(\mathbf{x})$ is the $\tilde{\alpha}$-*mixture density*. The $\tilde{\alpha}$-*relative density ratio* is defined as:

$$r_{\tilde{\alpha}}(\mathbf{x}) = \frac{p(\mathbf{x})}{p'_{\tilde{\alpha}}(\mathbf{x})} = \frac{p(\mathbf{x})}{\tilde{\alpha}\, p(\mathbf{x}) + (1-\tilde{\alpha})\, p'(\mathbf{x})}. \tag{4.86}$$

Notice that, for $\tilde{\alpha} = 0$, this expression reduces to the plain density ratio and it is bounded above by $1/\tilde{\alpha}$ for $\tilde{\alpha} > 0$. Since (4.85) is not symmetric, the following divergence has been proposed in [34]:

$$\mathrm{P}E_{\tilde{\alpha}}\left(P \,\|\, P'\right) + \mathrm{P}E_{\tilde{\alpha}}\left(P' \,\|\, P\right). \tag{4.87}$$

Computing the estimate of the density ratio. A possible model for (4.85) is:

$$g\left(\mathbf{x}; \boldsymbol{\theta}\right) \equiv \sum_{n=1}^{N} \theta_n \cdot K\left(\mathbf{x}, \mathbf{z}_n\right), \tag{4.88}$$

where $\boldsymbol{\theta} = [\theta_1, \ldots, \theta_N]^\top \in \mathbb{R}^{N\times 1}$ are the model parameters, $K(\cdot, \cdot)$ is a kernel basis function, and $\mathbf{z}_d$ refers to the d-th data sample in $\mathbf{Y}(t)$. A commonly used kernel is the Gaussian one:

$$K\left(\mathbf{z}_1, \mathbf{z}_2\right) = exp\left(-\frac{\|\mathbf{z}_1 - \mathbf{z}_2\|^2}{2\iota^2}\right), \tag{4.89}$$

where $\iota > 0$ denotes the width of the kernel. The parameters $\boldsymbol{\theta}$ are estimated by minimizing the following cost:

$$
\begin{aligned}
J(\boldsymbol{\theta}) &= \frac{1}{2}\int p'_{\tilde{\alpha}}(\mathbf{x})\Big(r_{\tilde{\alpha}}(\mathbf{x}) - g(\mathbf{X};\boldsymbol{\theta})\Big)^2\, d\mathbf{x} \\
&= \frac{1}{2}\int p'_{\tilde{\alpha}}(\mathbf{x}) r^2_{\tilde{\alpha}}(\mathbf{x})\, d\mathbf{x} - \int p(\mathbf{x}) g(\mathbf{x};\boldsymbol{\theta})\, d\mathbf{x} \\
&+ \frac{\tilde{\alpha}}{2}\int p(\mathbf{x}) g(\mathbf{x};\boldsymbol{\theta})^2\, d\mathbf{x} + \frac{1-\tilde{\alpha}}{2}\int p'(\mathbf{x}) g(\mathbf{x};\boldsymbol{\theta})^2\, d\mathbf{x}.
\end{aligned}
\tag{4.90}
$$

The first term of (4.90) can be discarded since it is independent by the unknown parameters. By adding a regularization term $\frac{\lambda}{2}\boldsymbol{\theta}^\top\boldsymbol{\theta}$, using the definition (4.88) in (4.90) and approximating the expectations with empirical averages, a minimization problem is obtained as

$$
\widehat{\boldsymbol{\theta}} = \underset{\boldsymbol{\theta}\in\mathbb{R}^N}{\arg\min}\left[\frac{1}{2}\boldsymbol{\theta}^\top\widehat{\mathbf{F}}\boldsymbol{\theta} - \widehat{\mathbf{f}}^\top\boldsymbol{\theta} + \frac{\lambda}{2}\boldsymbol{\theta}^\top\boldsymbol{\theta}\right],
\tag{4.91}
$$

where $\widehat{\mathbf{F}} \in \mathbb{R}^{N\times N}$, $\widehat{\mathbf{f}} \in \mathbb{R}^{N\times 1}$ and $\lambda > 0$ controls the regularization strength.

The (d, e)-element of $\widehat{\mathbf{F}}$ reads as

$$
\begin{aligned}
\widehat{F}_{(d,e)} &= \frac{\tilde{\alpha}}{N}\sum_{i=1}^{N} K(\mathbf{z}_i, \mathbf{z}_d)\cdot K(\mathbf{z}_l, \mathbf{z}_e) \\
&+ \frac{1-\tilde{\alpha}}{N}\sum_{j=1}^{N} K(\mathbf{z}'_j, \mathbf{z}_d)\cdot K(\mathbf{z}'_j, \mathbf{z}_e).
\end{aligned}
\tag{4.92}
$$

The l-th component of $\widehat{\mathbf{f}}$ reads as

$$
\widehat{h}_{(d)} = \frac{1}{N}\sum_{n=1}^{N} K(\mathbf{y}_n, \mathbf{y}_d).
\tag{4.93}
$$

The solution (4.91) can be computed as

$$
\widehat{\boldsymbol{\theta}} = \left(\widehat{\mathbf{F}} + \lambda I_N\right)^{-1}\cdot\widehat{\mathbf{f}},
\tag{4.94}
$$

Then, the density ratio estimator is $g\left(\mathbf{x};\hat{\boldsymbol{\theta}}\right) = \sum_{n=1}^{N}\widehat{\theta}_n \cdot K(\mathbf{x}, \mathbf{z}_n)$.

Computing the divergence. To use $g\left(\mathbf{x};\widehat{\boldsymbol{\theta}}\right)$, it is possible to rewrite the Pearson divergence (4.85) as

$$
\begin{aligned}
\mathrm{P}E_{\tilde{\alpha}}\left(P \,\|\, P'\right) &= \tfrac{1}{2}\int p'_{\tilde{\alpha}}(\mathbf{x}) \cdot \left(\frac{p(\mathbf{x})}{p'_{\tilde{\alpha}}(\mathbf{x})} - 1\right)^2 d\mathbf{x} \\
&= \frac{1}{2}\int p'_{\tilde{\alpha}}(\mathbf{x}) \cdot \left(\tfrac{p(\mathbf{x})^2}{p'_{\tilde{\alpha}}(\mathbf{x})^2} - 2\tfrac{p(\mathbf{x})}{p'_{\tilde{\alpha}}(\mathbf{x})} + 1\right) d\mathbf{x} \\
&= \frac{1}{2}\int \left(\tfrac{p(\mathbf{x})^2}{p'_{\tilde{\alpha}}(\mathbf{x})} - 2p(\mathbf{x}) + p'_{\tilde{\alpha}}(\mathbf{x})\right) d\mathbf{x} \\
&= \frac{1}{2}\int \left(\frac{p(\mathbf{x})}{p'_{\tilde{\alpha}}(\mathbf{x})}\right) \cdot p(\mathbf{x})\, d\mathbf{x} - \frac{1}{2}.
\end{aligned}
\tag{4.95}
$$

The approximation of (4.85) is then obtained by substituting $g\left(\mathbf{x}; \widehat{\boldsymbol{\theta}}\right)$ in (4.95)

$$
\widehat{\mathrm{PE}}_{\tilde{\alpha}} = \frac{1}{2N}\sum_{n=1}^{N} \widehat{g}\left(\mathbf{z}_n\right) - \frac{1}{2}. \tag{4.96}
$$

The final computed score (4.87) is the quantity $\pi \equiv \widehat{\mathrm{PE}}_{\tilde{\alpha}}(P \,\|\, P') + \widehat{\mathrm{PE}}_{\tilde{\alpha}}(P' \,\|\, P)$, that represents an indication of dissimilarity between the two data sets $\mathbf{Y}$ and $\mathbf{Y}'$.

4.5.2.2 Feature Computation for EMA Condition Monitoring

The change detection method based on the RuLSIF approach can be used with any dataset described by a set of features. In the specific case of EMA monitoring, we chose, as in the HOLMES project (see Sects. 4.2.1 and 4.5.1), to employ features computed from phase currents data, see Fig. 4.13.

As typical with controlled systems, the emerging of a fault can be hidden by the closed-loop system configuration: as such, monitoring the reference-output variables can be ineffective. On the contrary, monitoring the phase currents, i.e., the control actions of the system, can provide more diagnostic information.

Consider the phase currents measurements, $\mathbf{i}_q^{(i)} \in \mathbb{R}^{N_c \times 1}$ for $i = 1, \cdots, p$, where $p = 1, 2, 3$ is the number of current phases, acquired during an experiment q. Since, in the considered REPRISE setup, the acquired position evolves as a sinusoid with period T, it is possible to write the current signal of the p-th phase in the τ period as

$$
i^{(p)}(t), \quad (\tau - 1)T \leq t \leq \tau T, \tag{4.97}
$$

with $\tau = 1, \ldots, N_p$, being N_p the total number of periods in the considered experiment, such that the number of measured data is $N_c = N_p \cdot T$.

As simple, yet effective features that can be computed from phase currents, consider the Root Mean Square (RMS) and the Crest Factor (CF) indicators. Each of them is calculated for each phase current $p = 1, 2, 3$, by considering a single period $\tau = 1, \ldots, N_p$ as follows.

Root Mean Square indicator The RMS indicator in a period τ for the phase p is computed as

$$\sigma_p(\tau) = \sqrt{\frac{1}{T} \sum_{t=(\tau-1)T}^{\tau T} \left[i^{(p)}\right]^2(t)}. \tag{4.98}$$

By averaging the RMS over the phases, the average RMS in period τ is given by

$$\Sigma(\tau) = \frac{1}{3}\Big(\sigma_a(\tau) + \sigma_b(\tau) + \sigma_c(\tau)\Big). \tag{4.99}$$

Crest Factor indicator The CF indicator in a period τ for the phase p is computed as

$$\gamma_p(\tau) = \frac{max\left(|i^{(p)}(t)|\right)}{\sigma_p(\tau)}. \tag{4.100}$$

By averaging the CF over the phases, the average CF in period τ is given by $\tau = 1, \ldots, N_p$ as

$$\Gamma(\tau) = \frac{1}{3}\Big(\gamma_a(\tau) + \gamma_b(\tau) + \gamma_c(\tau)\Big). \tag{4.101}$$

These RMS and CF features are computed for different operating frequencies (see Sect. 4.2.2.2), that is $\{0.1, 0.3, 0.5, 0.8, 0.9, 1\}$ Hz. In what follows, results have been reported by considering an operating frequency of 1 Hz, using a position amplitude of 10 mm.

4.5.2.3 Batch Change Detection for EMA Condition Monitoring

The change detection approach presented in Sect. 4.5.2.1 is inherently *online*, i.e., it processes the data as they arrive. While this is a nice feature for a monitoring algorithm, it has to be noted that, given the knowledge-based nature of the method, it is advisable to control for as many external factors as possible. Examples include external loads or temperature variations.

Based on this rationale, it is possible to adapt the online change detection method previously described to work in a *batch* fashion, by comparing the data from one experiment to the data from another (past) one. The batch approach requires a reference experiment to be used for comparison with new ones. A strategy proposed in [41] is to compare the actual experiment with the last one, in which change detection score exceeded a certain value. The number of times that this threshold is exceeded, gives a score that can be used for monitoring purposes.

A total of $Q = 11$ experiments in the H0 (nominal) condition are considered, see Sect. 4.2.2.2 and Fig. 4.14. The tests of 11 and 18 Sept. 2017 are interpreted as experiments in healthy state. Tests of 21 and 25 Sept. 2017 were acquired after trials in H1 (overload) condition, where the partial lubricant is present. From this point on, tests without lubricant were performed.

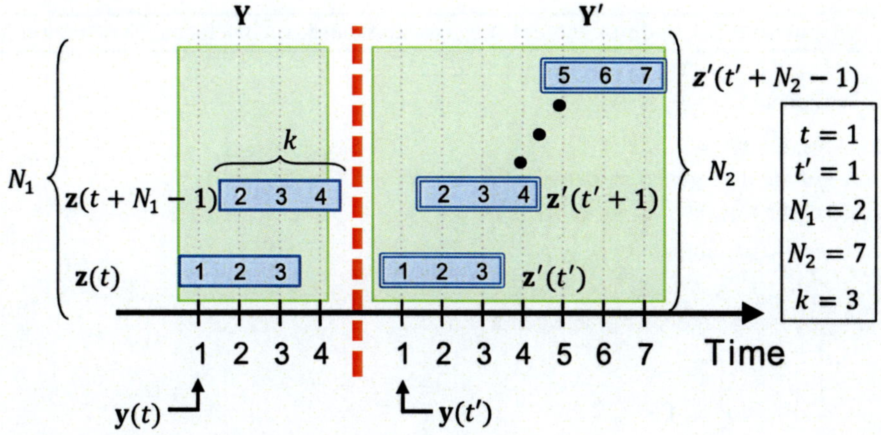

Fig. 4.71 Batch RuLSIF data management example, $m = 1$, $k = 3$, $N_1 = 2$ and $N_2 = 7$. © IEEE 2018. Reprinted, with permission, from [41]

With respect to the online version of change detection described in Sect. 4.5.2.1, the batch approach does not employ a sliding window: instead, a batch of N_1 data $\mathbf{Y}$ from one experiment is compared to a batch of N_2 data $\mathbf{Y}'$ from another one, see Fig. 4.71. The divergence is compared as previously described.

Since two features are computed (the RMS and the CF indicators), the time series has dimensionality $m = 2$. Each sample $\mathbf{y}(\tau) \in \mathbb{R}^{2\times 1}$ contains the elements $\Sigma(\tau)$ and $\Gamma(\tau)$. The hyperparameters of the method are set as in [41], i.e., $k = 1$, $\tilde{\alpha} = 0.5$, while ι and λ can be chosen via κ-fold cross-validation for every comparison, or set in advance and fixed for all comparisons. Being k related to the system memory, a value $k = 1$ means that the computed indicators are assumed to not depend on their past. Each test q is represented by a data matrix $\mathbf{Y}_q \in \mathbb{R}^{100\times 2}$, i.e., $m = 2$ features and $N_p = 100$ observations that correspond to the number of periods of the input, see Table 4.9.

The condition monitoring indication is computed as given in Algorithm 1 and reported in [42]. As a first step, the divergence π_1 between two healthy tests (i.e., 11 and 18 Sept. 2017) is computed, see 4.14. Then, a threshold $\xi = 2 \cdot \pi_1$ is set. The incoming experiments are compared with the last one $q = q^*$ which divergence score exceeded the value ξ: every time this happens, a damage score ζ_q is increased. The results, reported in Fig. 4.72, show the effectiveness of the approach, where also the computed features are represented along with the divergence score and damage counter indicators.

Algorithm 1: Condition monitoring algorithm based on batch change detection.

Input: $\mathbf{Y}_{q^*}$, $\mathbf{Y}_q$, hyperparameters k, ι, λ
Output: ζ_q

Design phase (for $t \leq 1$):
get the divergence π_1 between $\mathbf{Y}_0$ and $\mathbf{Y}_1$
set the threshold $\xi \leftarrow 2 \cdot \pi_1$

Test phase (for $t > 1$):
$\zeta_q \leftarrow 0$; $q^* \leftarrow 1$
get the divergence π_q between $\mathbf{Y}_{q^*}$ and $\mathbf{Y}_q$
if $\pi_q > \xi$ **then**
  $q^* \leftarrow q$
  $\zeta_q \leftarrow \zeta_q + 1$
end

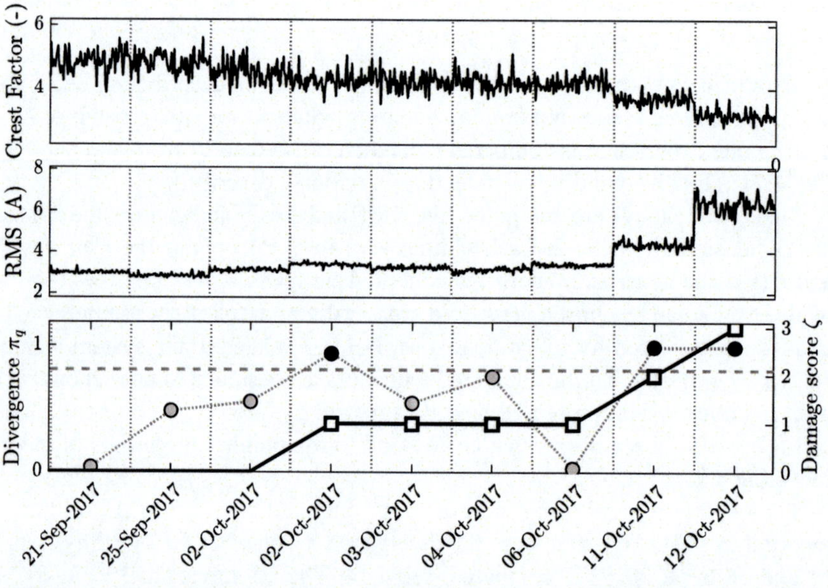

Fig. 4.72 Results of batch change detection algorithm for condition monitoring. (Top) Crest factor. (Middle) Root Mean Square. (Bottom) computed divergence π_q (gray and black dots) and monitoring score ζ_q (white squares). Reproduced from [40]—originally published open access and licensed under CC-BY 4.0. https://ieeexplore.ieee.org/stamp/stamp.jsp?tp=&arnumber=8878102

4.5.3 Knowledge-Based Condition Monitoring via Statistical Process Monitoring Techniques

This section presents the application of Statistical Process Monitoring (SPM) techniques for developing a condition monitoring approach for EMA of the REPRISE project of Sect. 4.2.2. We will focus on the *first phase* of the project (see Sects. 4.2.2.1 and 4.2.2.2).

4.5.3.1 Motivation of the Approach

Contrary to classic control charts that look at only one variable at a time, statistical process monitoring methods are able to leverage multivariate data for the aim of fault diagnosis [56]. The most famous indicators used in this sense are the Hotelling's T^2 statistic and the Q statistic, also known as the Squared Prediction Error (SPE), [50, 64]. Contribution and reconstruction plots have been devised with the purposes of fault isolation and identification [2, 28, 56].

Statistical process monitoring methods can be preferred to model-based approaches since, essentially, the design effort is lower. However, SPM approaches output a dichotomous answer, i.e., a fault vs. no fault indication. It is clear that this information is not suited to condition monitoring, since no indications about system degradation are given.

This section presents an approach for devising condition monitoring indicators leveraging the results of the application of statistical process monitoring techniques on the considered process. In the context of the first phase of the REPRISE project, the process is the EMA described in Sect. 4.2.2.1. The proposed method assumes the role of a *preflight test strategy*, i.e., a test that has to be performed on the ground before (or after) the aircraft departure (as described for the change-point detection method described in Sect. 4.5.2).

4.5.3.2 Introduction to Statistical Process Monitoring

Let $\mathbf{x} \in \mathbb{R}^{m\times 1}$ be a feature vector of m variables. The data matrix $\mathbf{X} \in \mathbb{R}^{N\times m}$ groups N samples as

$$\mathbf{X} = [\mathbf{x}(1) \quad \mathbf{x}(2) \quad \dots \quad \mathbf{x}(N)]^\top . \tag{4.102}$$

As first processing step, $\mathbf{X}$ is usually normalized such that each feature has zero mean and unit variance. The covariance of $\mathbf{x}$ can then be approximated as

$$\mathbf{S} = \frac{1}{N-1}\mathbf{X}^\top\mathbf{X}. \tag{4.103}$$

The Principal Components Analysis (PCA) algorithm is used to decompose $\mathbf{S}$ as the sum of two contributions:

$$\mathbf{S} = \mathbf{P}\Lambda\mathbf{P}^\top + \tilde{\mathbf{P}}\tilde{\Lambda}\tilde{\mathbf{P}}^\top = \hat{\mathbf{S}} + \tilde{\mathbf{S}}, \tag{4.104}$$

where $\hat{\mathbf{S}} = \mathbf{P}\Lambda\mathbf{P}^\top$ and $\tilde{\mathbf{S}} = \tilde{\mathbf{P}}\tilde{\Lambda}\tilde{\mathbf{P}}^\top$. Denote with $l < m$ the number of Principal Components (PCs) that are selected. Then, $\mathbf{P} \in \mathbb{R}^{m\times l}$ and $\tilde{\mathbf{P}} \in \mathbb{R}^{m\times (m-l)}$ contain the principal and residual *loading vectors*, respectively. The diagonal matrices $\Lambda \in \mathbb{R}^{l \times l}$ and $\tilde{\Lambda} \in \mathbb{R}^{(m-l)\times(m-l)}$ contain the *principal eigenvalues set* $\hat{\mathcal{S}} = \{\sigma_i\}_{i=1,\cdots,l}$ and the *residual eigenvalues set* $\tilde{\mathcal{S}} = \{\tilde{\sigma}_i\}_{i=l+1,\cdots,m}$, respectively.

Squared Prediction Error (SPE) index or Q statistic. The Q statistic (or SPE index) of a point $\mathbf{x}$ is defined as:

$$Q(\mathbf{x}) \equiv \left\|\left(\mathbf{I} - \mathbf{P}\mathbf{P}^\top\right) \cdot \mathbf{x}\right\|_2^2. \tag{4.105}$$

Given a confidence level $(1 - \alpha) \times 100\%$, it is possible to set a threshold for Q as $Q_\alpha = g^{SPE}\chi^2_\alpha\left(h^{SPE}\right)$ with $g^{SPE} = \frac{\theta_2}{\theta_1}$, $h^{SPE} = \frac{\theta_1^2}{\theta_2}$, $\theta_j = \sum_{i=l+1}^{m} \tilde{\sigma}_i^j$, where $\tilde{\sigma}_i \in \tilde{\mathcal{S}}$ is the i-th eigenvalue of $\mathbf{S}$, and $\chi^2_\alpha\left(h^{SPE}\right)$ is the h^{SPE} degrees of freedom Chi-squared distribution deviate that corresponds to the $(1 - \alpha)$ percentile [2, 56].

Hotelling's T^2 statistic. The T^2 statistic of a point $\mathbf{x}$ is defined as

$$T^2(\mathbf{x}) \equiv \mathbf{x}^\top\mathbf{P}\Lambda^{-1}\mathbf{P}^\top\mathbf{x} = \mathbf{x}^\top\mathbf{D}\mathbf{x}, \tag{4.106}$$

where $\mathbf{D} = \mathbf{P}\Lambda^{-1}\mathbf{P}^\top \in \mathbb{R}^{n\times n}$ is a positive semidefinite matrix. Given a confidence level $(1 - \alpha) \times 100\%$ and assumptions of Gaussianity of the data, it is possible to set a threshold for the T^2 as $T^2_\alpha = \frac{l(N-1)(N+1)}{N(N-l)} F_\alpha(l, N - l)$, where $F_\alpha(l, N - l)$ indicates the deviate corresponding to the $(1 - \alpha)$ percentile of a F-distribution with l and $N - l$ degrees of freedom [56].

Then, a point $\mathbf{x}$ is said to be in *nominal conditions* if $T^2(\mathbf{x}) \leq T^2_\alpha$; otherwise, the point belongs to an out of control condition.

Fault Detection using SPE and T^2 statistics. Like the change detection method presented in Sect. 4.5.2, fault detection based on Q and T^2 statistcs consists in a two-step procedure, summarized in Algorithm 2:

1. *design phase*: the thresholds, normalization quantities and projection matrices are defined using healthy data;
2. *test phase*: the algorithm is employed on unseen data.

Regarding the tuning of the hyperparameters of the method, the following guidelines can be followed:

- the α level determines the value of the threshold and so the number of detected violations;
- the number l of retained principal components determines the level of approximation of the data projection and can be tuned on the basis of the signal-to-noise (S/N) ratio in the measurements.

4.5.3.3 Condition Monitoring of EMAs Based on SPM Approaches

The proposed approach for the condition monitoring of EMA is based on the analysis of the phase currents during a preflight test. The main steps can be summarized as reported in Algorithm 3.

Algorithm 2: Fault detection based on Q and T^2 statistics.

Input: Training data $\mathbf{X}$ and test data $\mathbf{X}^*$, l, α
Output: Fault / no fault indication

Design phase:
1 normalize $\mathbf{X}$ to zero mean and unit variance
2 compute the matrices $\mathbf{P}$, Λ and $\tilde{\Lambda}$
3 compute the Q_α and T^2_α

Test phase:
4 normalize $\mathbf{X}^*$ using the same mean and variances used in step 1
5 compute the statistics (4.105) and (4.106) on the normalized test data using $\mathbf{P}$, Λ and $\tilde{\Lambda}$ computed in step 2
6 compare each statistic with its threshold: raise an alarm in case of violations

1. Denote with $\mathbf{i}_q^{(i)} \in \mathbb{R}^{N\times 1}$ for $i = 1, \cdots, p$ the vectors containing the i-th phase current measurements in the q-th experimental test. The first experiment $q = 1$ is used to compute the normalization quantities and the thresholds T^2_α and Q_α.
2. Construct the data matrix $\mathbf{X}_q \in \mathbb{R}^{N\times(p-1)}$ as

$$\mathbf{X}_q = \left[\mathbf{i}_q^{(1)} \;\; \mathbf{i}_q^{(2)} \cdots \mathbf{i}_q^{(p-1)}\right] = \begin{bmatrix} \mathbf{x}_q(1)^\top \\ \vdots \\ \mathbf{x}_q(N)^\top \end{bmatrix}. \tag{4.107}$$

   The variables $\mathbf{x}_q(j) \in \mathbb{R}^{(p-1)\times 1}$, $j = 1, \ldots, N$ are denoted in such a way that $\mathbf{x}_q(j) = \left[\mathbf{i}_q^{(1)}[j] \;\; \mathbf{i}_q^{(2)}[j] \cdots \mathbf{i}_q^{(p-1)}[j]\right]^\top$, and $\mathbf{i}_q^{(i)}[j]$ is the j-th component of the vector $\mathbf{i}_q^{(i)}$, $i \in \{1, \ldots, p-1\}$. Notice that only $p-1$ phase currents are needed, due to their linear dependence in BLDC motors (phase currents should sum to zero).
3. Perform PCA on $\mathbf{X}_q$, retaining a number of components l based on the S/N ratio or on the amount of data variance to be retained.
4. Compute the $Q_q\left(\mathbf{x}_q(j)\right)$ and $T^2_q(\mathbf{x}_q(j))$ values and compare them with the respective thresholds.
5. Each threshold violation defines an *event* $e_q(j)$ s.t. for $j = 1, \ldots, N$:

$$e_q(j) = \begin{cases} 1 \;\; if \;\; T^2_q(j) > T^2_\alpha \\ 0 \;\; if \;\; T^2_q(j) \le T^2_\alpha \end{cases} \tag{4.108}$$

   and similarly for the Q statistic.

The main assumption is that *an higher number of threshold violations is caused by a lower system health.* The number of events is used to distillate several health monitoring indicators that give the monitoring information.

1. **Event frequency**

   The *event frequency* indicator λ_q is defined as the percentage of observed events N_q over the number of data N in an observation time t_e with sampling time T_s, s.t. $N = t_e \cdot 1/T_s$ and $N_q = \sum_{j=1}^{N} e_q(j)$:

$$\lambda_q = \frac{\text{\# of events}}{\text{\# of data}} \cdot 100\% \equiv \frac{N_q}{N} \cdot 100. \tag{4.109}$$

2. **Mean time to event**

   The *mean time to event* (MTTE) indicator Δ_q computes the average time that lasts between any two consecutive events. We call these quantities *inter-arrival times*. With N_q events, there are $N_q - 1$ inter-arrival times. Define with $K \subset \mathbb{N}$ the ordered set (in increasing order) of indices j s.t. $e_q(j) = 1$. Let the number of observations between event i and event $i+1$, with $i = 1, \ldots, N_q - 1$, be $\delta_{i,q} = (k_{i+1} - k_i)$, with $k_i \in K$ the index when event i occurs. The indicator Δ_q is, therefore, computed as

$$\Delta_q = \frac{\text{sum of inter-arrival times}}{\text{\#of inter-arrival times}} \equiv \frac{\sum_{i=1}^{N_q-1} \delta_{i,q}}{N_q - 1} \cdot T_s. \tag{4.110}$$

3.-4. **Weibull distribution of inter-arrival times rising edges**

The Weibull distribution is a vastly employed tool in reliability engineering and survival analysis for modeling random variables that represent *times* [27].

Algorithm 3: Condition monitoring based on statistical process monitoring.

Input: $\mathbf{i}_q^{(1)}, \mathbf{i}_q^{(2)}, \ldots, \mathbf{i}_q^{(p-1)}, l, \alpha$
Output: q-th value of the monitoring indicators

1 compute $\mathbf{X}_q$ in (4.107)

Design phase (for $q = 1$):

2 run steps 1-2-3 of Algorithm 2($\mathbf{X}_1, l, \alpha$) to compute $Q_\alpha, T_\alpha^2, \mathbf{P}, \Lambda, \tilde{\Lambda}$

Training or setup phase (for $q > 0$):

3 run steps 4-5 of Algorithm 2($\mathbf{X}_q, l, \alpha$) to compute $e_q(j)$ in (4.108), for $j = 1, \ldots, N$

4 compute $\lambda_q, \Delta_q, \hat{\eta}_q, \hat{\beta}_q$

In particular, when the modeled variables consist of "time-to-failure" data, the Weibull distribution is used to model the *failure rate* of the components subject to failure, see Sect. 2.1.

As the considered *component degrades*, we expect that:

- the average time-to-failure *gets lower*, since we expect an impending failure;
- the standard deviation of the times-to-failure *gets lower*, since we are more confident about the imminent event.

Define as *rising edge* the first event $e_q(j)$ of a continuous strike of consecutive events. Then, store the times between consecutive rising edges in the variables $r_q(z)$, $z = 1, \ldots, R_q$ with R_q the number of rising edges in experimental test q. Values in $r_q(z)$ are called *time-to-failure* data. Thus, it is possible to fit a Weibull distribution via maximum-likelihood to this data, and use the estimated parameters $\hat{\eta}_q$ and $\hat{\beta}_q$ has monitoring indicators. The same result applies if falling edges are used to define time-to-failure data. When the observation time ends before observing the event of interest, a *right-censored* time-to-failure value is present: this situation can be managed by properly considering the Weibull likelihood to be used.

It is now interesting to ask about the interpretation of these indices. The event frequency index $\hat{\lambda}_q$ and the meantime to event Δ_q have a simple interpretation as "how many events per unit of time" or "how fast the events are happening", respectively. In the former case, a threshold can be put at a value of, e.g., 50 %. In the latter case, a natural limit is the sampling time T_s, and a threshold can be set at $2T_s$. The Weibull shape parameter $\hat{\beta}_q$ has an intuitive interpretation: if greater than 1, the failure rate is increasing with time; on the contrary, the scale parameter $\hat{\eta}_q$ is less intuitive, and a threshold on its relative variation can be employed.

The indicators (4.109) and (4.110) are very simple to compute, while accurate estimation of the Weibull parameters requires an optimization process. The choice about which indicators to use depend on the application field and the safety requirements.

4.5.3.4 Results on the REPRISE Phase 1 EMA

The application of the proposed approach on the phase 1 of the REPRISE project relied on data from Monitoring trials (21 Sept. 2017–12 Oct. 2017), see Fig. 4.14, so that $Q = 9$ experiments are employed. The dataset of 18 Sept. 2017, is used as the healthy $q = 1$ dataset to design the thresholds and to compute the normalization means and variances, using $\alpha = 0.05$.

With particular reference to Monitoring trials as described in Sect. 4.2.2.2, the subset of data consisting in tests at 0 m*m* offset and 10 m*m* amplitude are employed. This choice approximates the motor usage described in Table 4.8, and it is also a condition that can be easily performed in a periodic preflight or maintenance test.

The matrix $\mathbf{X}_q \in \mathbb{R}^{N \times 2}$, representing the result of the q-th Monitoring trial, is computed as follows for $q = 1, \ldots, Q$:

$$\mathbf{X}_q = \left[\mathbf{i}_q^{(2)} \ \ \mathbf{i}_q^{(3)}\right], \tag{4.111}$$

where the quantities $\mathbf{i}_q^{(2)}, \mathbf{i}_q^{(3)}$ in (4.111) are defined by considering all the currents measurements at different position frequencies. Let $F = \{f_1, f_2, \ldots, f_{n_f}\}$ be a set of n_f frequencies. Denote with $\mathbf{i}_{f,q}^{(b)}, \mathbf{i}_{f,q}^{(c)} \in \mathbb{R}^{N\times 1}$ the vectors containing the phase currents measurements in the q-th Monitoring trial, at a frequency f of the sinusoidal reference position profile s.t. $f \in F$, for the two motor phases 2, 3, respectively. The quantities $\mathbf{i}_q^{(p)} \in \mathbb{R}^{N\times 1}$, $p \in \{2, 3\}$ are computed as the sum of the two phase currents across all considered set of frequencies F

$$\mathbf{i}_q^{(p)} = \sum_{f\in F} \mathbf{i}_{f,q}^{(p)}. \tag{4.112}$$

The set of considered frequencies in (4.112) is (in Hz) $F = \{0.3, 0.5, 0.8, 0.9, 1\}$. This is a subset of $n_f = 5$ frequencies from the 10 frequencies used to perform Endurance and Monitoring trials, see Sect. 4.2.2.2.

In what follows, the events $e_q(j)$, $j = 1, \ldots, N$ are evaluated using only the T^2 statistics, because the columns of the matrices $\mathbf{X}_q$, $q = 1, \ldots, Q$, are found to be linearly independent, so that $l = m = 2$, and the Q statistic can not be used.

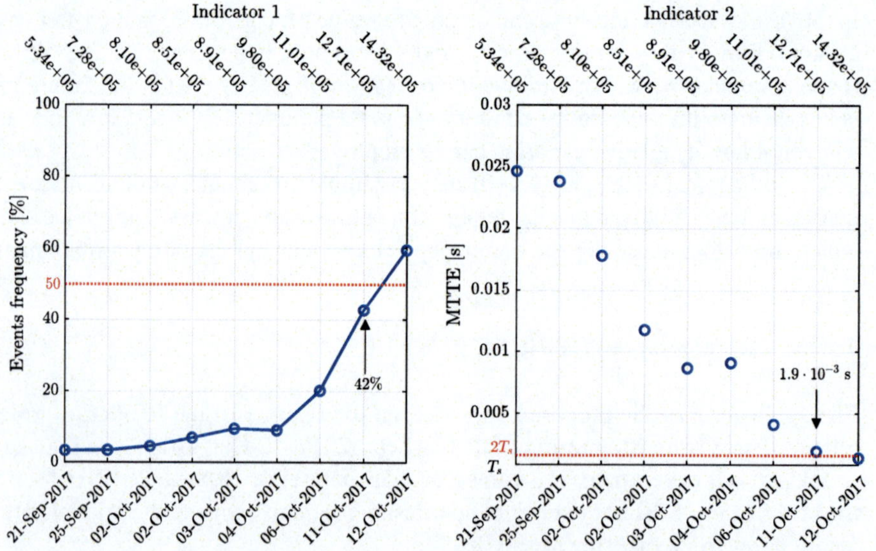

Fig. 4.73 Event frequency λ_q and MTTE Δ_q computed on $Q = 9$ Monitoring trials (bottom axis) and total number of screw revolutions after anti-rotation removal (top axis). Reproduced from [40]—originally published open access and licensed under CC-BY 4.0. https://ieeexplore.ieee.org/stamp/stamp.jsp?tp=&arnumber=8878102

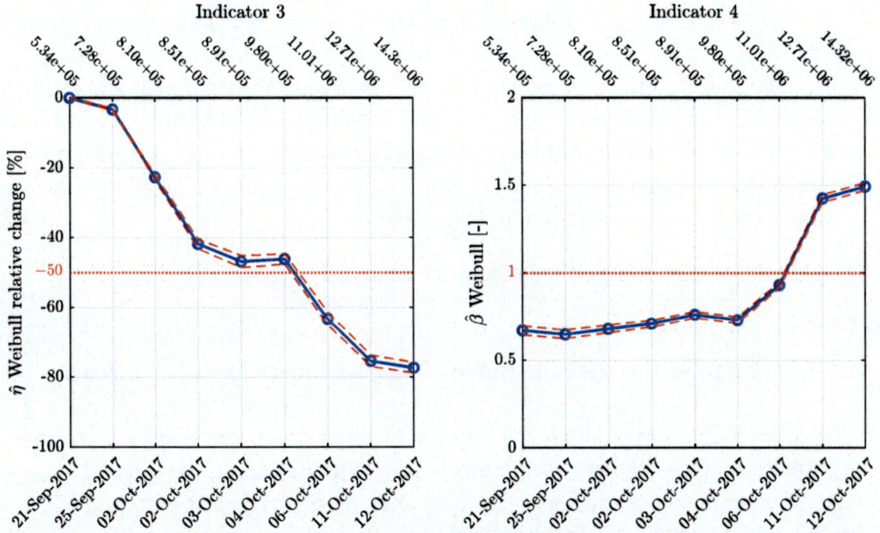

Fig. 4.74 Weibull scale parameter $\hat{\eta}_q$ (percentage change with respect to first value) and Weibull shape parameter $\hat{\beta}_q$ computed on Monitoring trials (bottom axis) and total number of screw revolutions after anti-rotation removal (top axis). Confidence intervals on estimates are shown (dashed red lines). Reproduced from [40]—originally published open access and licensed under CC-BY 4.0. https://ieeexplore.ieee.org/stamp/stamp.jsp?tp=&arnumber=8878102

The results of the computed indicator values are reported in Figs. 4.73 and 4.74, where a progressive trend can be noticed, in line with the actuator degradation. In particular:

- The *Event frequency* index λ_q increases exponentially with the degradation of the actuator, showing that the events are getting more frequent. In case of complete failure, the 100% value is reached.
- The *MTTE* index Δ_q decreases with the degradation, showing that the time between events is getting shorter. In case of complete failure, the MTTE will be equal to T_s. By the end of the tests, approximately 1 in every 10 measurements generates an event.
- Along with the *Weibull shape parameter* $\hat{\beta}_q$ and the *Weibull scale parameter* $\hat{\eta}_q$, it is possible to depict also the 95% confidence intervals on the parameters estimates. The index $\hat{\eta}_q$ is reported as percentage change with respect to the value $\hat{\eta}_1$. Since more events are present with higher degradation, more time-to-failure observations $r_q(z)$ are available. This translates into more reliable estimates with narrower confidence intervals. A warning threshold for $\hat{\beta}_q$ can be automatically set when $\hat{\beta}_q > 1$ (i.e., the failure rate increases with time). Figure 4.74 shows how this value is approached by the test of 06 Oct. 2017 and exceeded by the test of 11 Oct. 2017, in the same way as with the other indicators. Indicator $\hat{\eta}_q$ depends in a nontrivial way from data, so it is more complicated to set a threshold. In this case, an alarm can be issued when its value reaches, e.g., less than 50% of its initial one

$\hat{\eta}_1 = 0.044$, resulting in a threshold of 0.022. Again, this is true for the test of 11 Oct. 2017.

- The effect of $\hat{\eta}_q$ and $\hat{\beta}_q$ is shown in 4.76, that depicts the empirical and estimated survival functions $S\left(x|\beta_q, \eta_q\right)$, see (2.4). It can be seen how the mean value and the standard deviation of the estimated models decrease with degradation. This means that the times between rising edges are getting shorter and shorter, and there is less and less uncertainty about their value.

A synthesis of the proposed indicators for each test is reported in Fig. 4.75.

4.5.3.5 Comparison with the Batch Change-Point Detection Approach

As can be seen in Figs. 4.73, 4.72 and 4.74, both approaches detect a consistent degradation prior to the end of the motor life. With respect to the method based on change-point detection of Sect. 4.5.2, the indicators developed based on SPM are more intuitive. Thus, also the threshold definition task is easier.

The approach based on SPM indicators requires to define only two hyperparameters: (i) the confidence level α and the number of principal components l (or a percentage of explained variance). The definition of the first hyperparameter influence the discovery rates of the algorithm, while the definition of the second hyperparameter defines how much data variability is taken into account for diagnosis, and standard methods exist to make this choice.

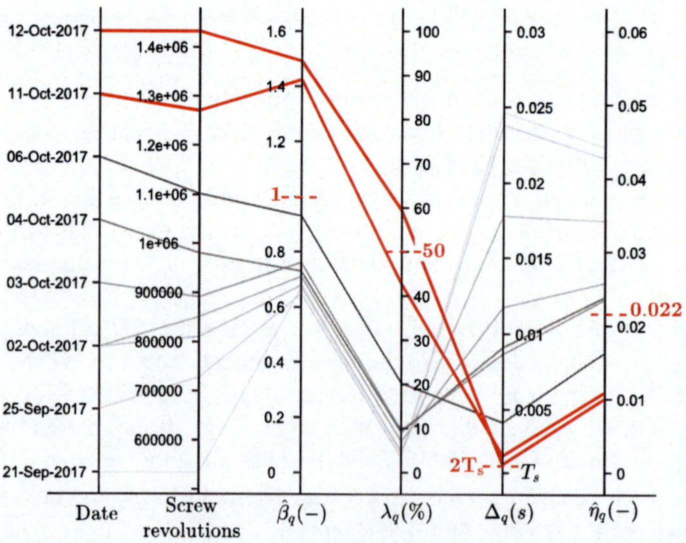

Fig. 4.75 Dates, screw revolutions, and indicators computed on the different tests. The line color ranges from earlier healthy tests (light gray) to later tests in degraded condition (red)

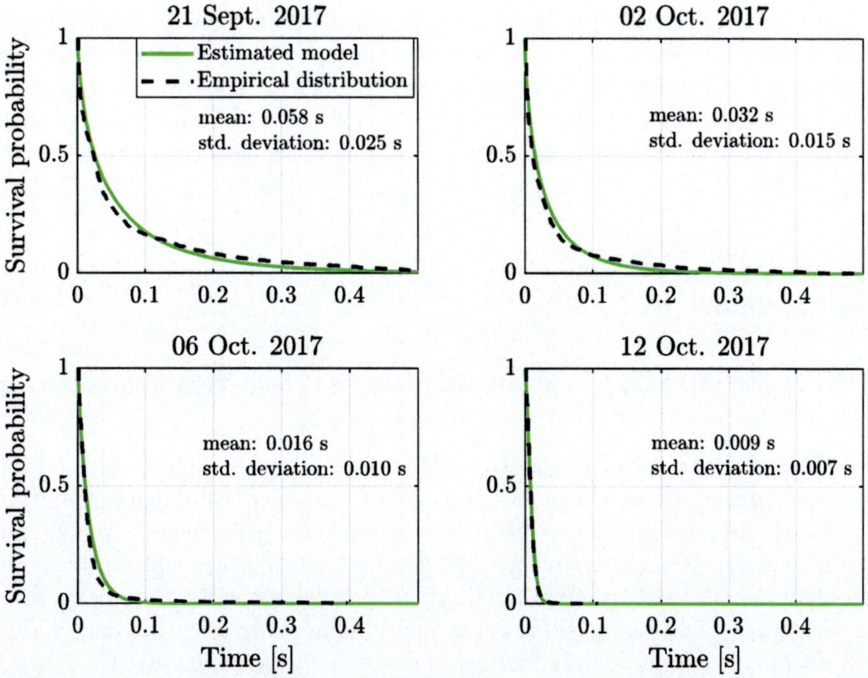

Fig. 4.76 Estimated Weibull distributions (green line) and empirical distribution from data (dashed black line), for different experimental tests

The change-point detection method requires the tuning of three hyperparameters: (i) the $\tilde{\alpha}$ term for computing the $\tilde{\alpha}$-relative Pearson divergence; (ii) a regularization term μ and (iii) the standard deviation ι of the Gaussian kernel for estimating the density ratio. Cross-validation procedures can be used to tune these knobs, but their interpretation is somewhat more difficult with respect to the SPM based approach.

The batch change detection method appears to be less computationally demanding and easier to deploy on an ECU, if the hyperparameters are set in advance and fixed for all the comparisons. The rationale based on SPM requires to store the projection matrices in order to compute the SPE and T^2 statistics, which can require a large ECU memory non easily available in flight applications.

The specific applicative case considered, the approach based on SPM takes into account more frequencies of motion (the reference position is sinusoidal), while change-point detection based method has to rely on only a single input frequency.

4.5.4 Final Considerations on Knowledge-Based Approaches

Knowledge-based approaches are widely applicable due to their "freedom" from prior knowledge and assumptions. The cons of using knowledge-based approaches lies in the fact that their applicability is limited on data that share the same characteristics of the data that are used to train those algorithms. Thus, varying operating conditions and external disturbances can compromise their effectiveness. For this reason, they are best suited as preflight tests, or in conditions, where external disturbances are kept controlled. Nonetheless, they can be used in conjunction with model-based approaches, that take care of disturbance rejection and adapt to dynamic conditions.

4.6 Summary

This chapter presented several practical examples of fault diagnosis and condition monitoring methods for airborne EMA. Summarizing:

- **Model-based approaches** are best suited when it is possible to develop (also by system identification techniques) a model of the system subjected to faults (and enough time and cost resources to develop and validate the model). Model-based approaches allow to design a residual generator which is decoupled from system inputs (control and disturbance actions), while minimizing the effect of the noises on the generated residuals. They also allow to simulate the effect of a fault, without resorting to a test bench for fault injection or fault degradation trials.
- **Signal-based approaches** perform well the signature that a fault lies in a specific set of signals is known a priori, and it is difficult to devise a model of the system (as in the case of bearing vibrations). However, they are not robust against external events.
- **Knowledge-based approaches** are the right tool when no prior information is available on the system. In this case, a large experimental campaign has to be performed to collect data in different operating regimes, in both healthy and not healthy configurations. These approaches, not being robust to external conditions, are best suited as test when the aircraft is at rest.

All the approaches can be used to devise fault diagnosis, condition monitoring, and fault prognosis solutions. Depending on the EMA part/component to be monitored, a monitoring approach can be preferable to another. Surely, the best solution would be the combination of two or more approaches, with the model-based technique that takes care of generating robust residuals, and, for example, the knowledge-based one that operates on those residuals (which are in some sort standardized, since they have not been affected by external events) in order to obtain more information about the faults (i.e., for isolation of identification purposes).

References

1. Reliability prediction of electronic equipment (1995). https://snebulos.mit.edu/projects/reference/MIL-STD/MIL-HDBK-217F-Notice2.pdf

2. Alcala CF, Qin SJ (2009) Reconstruction-based contribution for process monitoring. Automatica 45(7):1593–1600. https://doi.org/10.1016/j.automatica.2009.02.027
3. Arriola D, Thielecke F (2017) Model-based design and experimental verification of a monitoring concept for an active-active electromechanical aileron actuation system. Mech Syst Signal Process 94:322–345. https://doi.org/10.1016/j.ymssp.2017.02.039
4. Balaban E, Bansal P, Stoelting P, Saxena A, Goebel KF, Curran S (2009) A diagnostic approach for electro-mechanical actuators in aerospace systems. In: 2009 IEEE Aerospace conference, pp. 1–13. IEEE. https://doi.org/10.1109/AERO.2009.4839661
5. Benbouzid MEH (2000) A review of induction motors signature analysis as a medium for faults detection. IEEE Trans Ind Electron 47(5):984–993. https://doi.org/10.1109/41.873206
6. Brandt A (2011) Noise and vibration analysis: signal analysis and experimental procedures. John Wiley & Sons
7. Cameron J, Thomson W, Dow A (1986) Vibration and current monitoring for detecting airgap eccentricity in large induction motors. In: IEEE Proceedings B (Electric Power Applications), vol 133. IET, pp 155–163. https://doi.org/10.1049/ip-b.1986.0022
8. Cawley GC, Talbot NL (2010) On over-fitting in model selection and subsequent selection bias in performance evaluation. J Mach Learn Res 11(Jul):2079–2107
9. Cologni AL, Mazzoleni M, Previdi F (2016) Modeling and identification of an electro-hydraulic actuator. In: 2016 12th IEEE International Conference on Control and Automation (ICCA), pp 335–340. https://doi.org/10.1109/ICCA.2016.7505299
10. Combastel C, Lesecq S, Petropol S, Gentil S (2002) Model-based and wavelet approaches to induction motor on-line fault detection. Control Eng Pract 10(5):493–509. https://doi.org/10.1016/S0967-0661(01)00158-7
11. Denti E, Di Rito G, Galatolo R (2006) Real-time hardware-in-the-loop simulation of fly-by-wire flight control systems. In: 25th Congress of the International Council of the Aeronautical Sciences (ICAS). Hamburg, Germany, pp 3574–3581
12. Denti E, Di Rito G, Galatolo R, Schettini F (2011) Power absorption characterisation of electro-mechanical flight control actuators via detailed system modelling. In: Proceedings of 3rd CEAS Air&Space Conference-21st AIDAA Congress, Venezia, Italy, pp 1967–1973
13. Di Rito G, Denti E, Galatolo R (2008) Development and experimental validation of real-time executable models of primary fly-by-wire actuators. Proc Inst Mech Eng Part I J Syst Control Eng 222(6):523–542. https://doi.org/10.1243/09596518JSCE546
14. Di Rito G, Galatolo R, Schettini F (2016) Experimental and simulation study of the dynamics of an electro-mechanical landing gear actuator. In: 30th Congress of the International Council of the Aeronautical Sciences (ICAS), Daejeon, South Korea
15. Di Rito G, Galatolo R, Schettini F(2016) Self-monitoring electro-mechanical actuator for medium altitude long endurance unmanned aerial vehicle flight controls. Adv Mech Eng 8(5). https://doi.org/10.1177/1687814016644576
16. Di Rito G, Schettini F (2018) Health monitoring of electromechanical flight actuators via position-tracking predictive models. Adv Mech Eng 10(4). https://doi.org/10.1177/1687814018768146
17. Di Rito G, Schettini F, Galatolo R (2018) Model-based prognostic health-management algorithms for the freeplay identification in electromechanical flight control actuators. In: 2018 5th IEEE International Workshop on Metrology for AeroSpace (MetroAeroSpace). IEEE, pp 340–345. https://doi.org/10.1109/MetroAeroSpace.2018.8453552
18. Ding SX (2013) Model-based fault diagnosis techniques: design schemes, algorithms, and tools, 2nd edn. Springer Publishing Company, Incorporated
19. Ding SX (2014) Data-driven design of fault diagnosis and fault-tolerant control systems, 1st edn. Springer-Verlag, London
20. Filippetti F, Franceschini G, Tassoni C, Vas P (1998) Ai techniques in induction machines diagnosis including the speed ripple effect. IEEE Trans Indust Appl 34(1):98–108. https://doi.org/10.1109/28.658729
21. Fu J, Maré JC, Fu Y (2017) Modelling and simulation of flight control electromechanical actuators with special focus on model architecting, multidisciplinary effects and power flows. Chin J Aeronaut 30(1):47–65. https://doi.org/10.1016/j.cja.2016.07.006

22. Gibbs AL, Su FE (2002) On choosing and bounding probability metrics. Int Stat Rev 70(3):419–435. https://doi.org/10.1111/j.1751-5823.2002.tb00178.x
23. Goupil P (2010) Oscillatory failure case detection in the A380 electrical flight control system by analytical redundancy. Control Eng Pract 18(9):1110–1119. https://doi.org/10.1016/j.conengprac.2009.04.003
24. Henao H, Capolino G, Fernandez-Cabanas M, Filippetti F, Bruzzese C, Strangas E, Pusca R, Estima J, Riera-Guasp M, Hedayati-Kia S (2014) Trends in fault diagnosis for electrical machines: a review of diagnostic techniques. IEEE Indust Electron Mag 8(2):31–42. https://doi.org/10.1109/MIE.2013.2287651
25. Henao H, Demian C, Capolino GA (2003) A frequency-domain detection of stator winding faults in induction machines using an external flux sensor. IEEE Trans Indust Appl 39(5):1272–1279. https://doi.org/10.1109/IAS.2002.1043735
26. Isturiz A, Vinals J, Fernandez S, Basagoiti R, Torre Arnanz Edl, Novo J (2010) Development of an aeronautical electromechanical actuator with real time health monitoring capability
27. Jiang R, Murthy D (2011) A study of weibull shape parameter: properties and significance. Reliab Eng Syst Saf 96(12):1619–1626. https://doi.org/10.1016/j.ress.2011.09.003
28. Joe Qin S (2003) Statistical process monitoring: basics and beyond. J Chemom J Chemom Soc 17(8–9):480–502. https://doi.org/10.1002/cem.800
29. Kanamori T, Hido S, Sugiyama M (2009) A least-squares approach to direct importance estimation. J Mach Learn Res 10(Jul):1391–1445
30. Kawahara Y, Yairi T, Machida K (2007) Change-point detection in time-series data based on subspace identification. In: Data Mining, 2007. ICDM 2007. Seventh IEEE International Conference on. IEEE, pp 559–564. https://doi.org/10.1109/ICDM.2007.78
31. Khalak A, Goebel K (2008) Real-time probabilistic forecasting of wear degradation using a macro-scale physical model. In: 2008 IEEE Aerospace Conference, pp 1–8. https://doi.org/10.1109/AERO.2008.4526628
32. Lee HW, Kim TH, Choi C (2005) A novel internal fault analysis of a brushless dc motor using winding function theory. In: IEEE International Conference on Electric Machines and Drives. IEEE, pp 11–16. https://doi.org/10.1109/IEMDC.2005.195694
33. Lei Y, Zuo MJ, He Z, Zi Y (2010) A multidimensional hybrid intelligent method for gear fault diagnosis. Expert Syst Appl 37(2):1419–1430
34. Liu S, Yamada M, Collier N, Sugiyama M (2013) Change-point detection in time-series data by relative density-ratio estimation. Neural Netw 43:72–83. https://doi.org/10.1016/j.neunet.2013.01.012
35. Ljung L (1998) System identification: theory for the user. Pearson Education
36. Maier JF, Eckert CM, Clarkson PJ (2017) Model granularity in engineering design-concepts and framework. Design Science 3: https://doi.org/10.1017/dsj.2016.16
37. Mazzoleni M, Formentin S, Previdi F, Savaresi SM (2014) Fault detection via modified principal direction divisive partitioning and application to aerospace electro-mechanical actuators. In: 53rd IEEE Conference on Decision and Control, pp 5770–5775. https://doi.org/10.1109/CDC.2014.7040292
38. Mazzoleni M, Maccarana Y, Previdi F (2017) A comparison of data-driven fault detection methods with application to aerospace electro-mechanical actuators. IFAC-PapersOnLine 50(1):12797–12802. https://doi.org/10.1016/j.ifacol.2017.08.1837 20th IFAC World Congress
39. Mazzoleni M, Maroni G, Maccarana Y, Formentin S, Previdi F (2017) Fault detection in airliner electro-mechanical actuators via hybrid particle filtering. IFAC-PapersOnLine 50(1):2860–2865. https://doi.org/10.1016/j.ifacol.2017.08.640 20th IFAC World Congress
40. Mazzoleni M, Previdi F, Scandella M, Pispola G (2019) Experimental development of a health monitoring method for electro-mechanical actuators of flight control primary surfaces in more electric aircrafts. IEEE Access 7, 153,618–153,634. https://doi.org/10.1109/ACCESS.2019.2948781
41. Mazzoleni M, Scandella M, Maccarana Y, Previdi F, Pispola G, Porzi N (2018) Condition assessment of electro-mechanical actuators for aerospace using relative density-ratio estimation. IFAC-PapersOnLine 51(15):957–962. https://doi.org/10.1016/j.ifacol.2018.09.070 18th IFAC Symposium on System Identification SYSID 2018

42. Mazzoleni M, Scandella M, Maccarana Y, Previdi F, Pispola G, Porzi N (2018) Condition monitoring of electro-mechanical actuators for aerospace using batch change detection algorithms. In: 2018 IEEE Conference on Control Technology and Applications (CCTA), pp 1747–1752. https://doi.org/10.1109/CCTA.2018.8511334
43. Mazzoleni M, Scandella M, Previdi F, Pispola G (2020) Data on the first endurance activity of a brushless dc motor for aerospace applications. Data in Brief 29(105):153. https://doi.org/10.1016/j.dib.2020.105153
44. Merzouki R, Davila J, Fridman L, Cadiou J (2007) Backlash phenomenon observation and identification in electromechanical system. Control Eng Pract 15(4):447–457. https://doi.org/10.1016/j.conengprac.2006.09.001
45. Márton L, Lantos B (2009) Control of mechanical systems with stribeck friction and backlash. Syst Cont Lett 58(2):141–147. https://doi.org/10.1016/j.sysconle.2008.10.001
46. Nandi S, Toliyat HA, Li X (2005) Condition monitoring and fault diagnosis of electrical motors-a review. IEEE Trans Energy Convers 20(4):719–729. https://doi.org/10.1109/TEC.2005.847955
47. Ossmann D, Varga A (2015) Detection and identification of loss of efficiency faults of flight actuators. Int J Appl Math Comput Sci 25(1):53–63. https://doi.org/10.1515/amcs-2015-0004
48. Pearson K (1992) On the criterion that a given system of deviations from the probable in the case of a correlated system of variables is such that it can be reasonably supposed to have arisen from random sampling. In: Breakthroughs in Statistics. Springer, pp 11–28
49. Powers DM (2011) Evaluation: from precision, recall and f-measure to roc, informedness, markedness and correlation
50. Qin SJ (2012) Survey on data-driven industrial process monitoring and diagnosis. Annu Rev Control 36(2):220–234. https://doi.org/10.1016/j.arcontrol.2012.09.004
51. Randall RB (2011) Vibration-based condition monitoring: industrial, aerospace and automotive applications. John Wiley & Sons
52. Randall RB, Antoni J (2011) Rolling element bearing diagnostics-a tutorial. Mech Syst Signal Process 25(2):485–520. https://doi.org/10.1016/j.ymssp.2010.07.017
53. Rauber TW, do Nascimento EM, Wandekokem ED, Varejão FM, (2010) Pattern recognition based fault diagnosis in industrial processes: review and application. INTECH Open Access Publisher. https://doi.org/10.5772/9365
54. Rousseeuw PJ, Croux C (1992) Explicit scale estimators with high breakdown point. L1-Stat Anal Relat Methods 1, 77–9
55. RTCA DO-160G, Environmental Conditions and Test Procedures for Airborne Equipment (2010)
56. Russell EL, Chiang LH, Braatz RD (2012) Data-driven methods for fault detection and diagnosis in chemical processes. Springer Science & Business Media. https://doi.org/10.1007/978-1-4471-0409-4
57. Schoen RR, Habetler TG, Kamran F, Bartfield RG (1995) Motor bearing damage detection using stator current monitoring. IEEE Trans Indust Appl 31(6):1274–1279. https://doi.org/10.1109/IAS.1994.345491
58. Siyu C, Jinyuan T, Caiwang L, Qibo W (2011) Nonlinear dynamic characteristics of geared rotor bearing systems with dynamic backlash and friction. Mech Mach Theory 46(4):466–478. https://doi.org/10.1016/j.mechmachtheory.2010.11.016
59. Smith MJ, Byington CS, Watson MJ, Bharadwaj S, Swerdon G, Goebel K, Balaban E (2009) Experimental and analytical development of health management for electro-mechanical actuators. In: 2009 IEEE Aerospace conference, pp 1–14. https://doi.org/10.1109/AERO.2009.4839660
60. Stone G, Kapler J (1998) Stator winding monitoring. IEEE Indust Appl Mag 4(5):15–20. https://doi.org/10.1109/2943.715501
61. Sugiyama M, Suzuki T, Kanamori T (2012) Density ratio estimation in machine learning. Cambridge University Press. https://doi.org/10.1017/CBO9781139035613
62. Tang L, Kacprzynski GJ, Goebel K, Vachtsevanos G (2009) Methodologies for uncertainty management in prognostics. In: 2009 IEEE Aerospace conference, pp 1–12. https://doi.org/10.1109/AERO.2009.4839668

63. Vörös J (2010) Modeling and identification of systems with backlash. Automatica 46(2):369–374. https://doi.org/10.1016/j.automatica.2009.11.005
64. Yin S, Ding SX, Xie X, Luo H (2014) A review on basic data-driven approaches for industrial process monitoring. IEEE Trans Indust Electron 61(11):6418–6428. https://doi.org/10.1109/TIE.2014.2301773
65. Yu ZY, Niu T, Dong HL (2018) A jam-tolerant electromechanical system. In: ACTUATOR 2018; 16th International Conference on New Actuators, pp 1–4
66. Zarei J (2012) Induction motors bearing fault detection using pattern recognition techniques. Expert Syst Appl 39(1):68–73. https://doi.org/10.1016/j.eswa.2011.06.042
67. Zolghadri A (2017) Turning model-based fdir theory into practice for aerospace and flight-critical systems. https://doi.org/10.13009/EUCASS2017-692
68. Zolghadri A, Cieslak J, Efimov D, Henry D, Goupil P, Dayre R, Gheorghe A, Leberre H (2015) Signal and model-based fault detection for aircraft systems. IFAC-PapersOnLine 48(21), 1096–1101 (2015). https://doi.org/10.1016/j.ifacol.2015.09.673 9th IFAC Symposium on Fault Detection, Supervision and Safety for Technical Processes (SAFEPROCESS)

Chapter 5
Concluding Remarks

Outline of the chapter. This chapter is organized as follows. Section 5.1 summarizes the main contents of the book.

Section 5.2 presents some notes for practitioners for facing a fault diagnosis project, collected by the authors during the presented works. The takeaways are quite general and can be applied also in other applicative domains with respect to that of electro-mechanical actuators.

5.1 Fault Diagnosis for More Electric Actuation Technologies

The aim of this book is to discuss, motivate, and present fault diagnosis and condition monitoring applications for Electro-mechanical Actuators (EMAs) operating in aerospace applications. EMAs are foreseen to be employed not only for primary and secondary flight control surfaces, but also for landing gears, nose-wheel, brakes, winglets, and thrust vectoring control.

Chapter 1 presented a broad overview of the More Electric Aircraft (MEA) concept, discussing technological, environmental, societal, and market points of view. The EMA technologies are introduced and presented, with reference to their past and current employement in aircraft solutions. The motivation for the MEA is presented by summarizing the efforts of many research programs at European and international levels.

Specifically, the whole EMA architecture is reviewed, i.e., the control schemes, motor, mechanical transmission, sensors, and power electronics, with attention to hardware redundancy. Chapter 2 introduced the Reliability, Availability, Maintainability, and Safety (RAMS) analysis for aircraft applications. The systematic approach to System Safety Assessment (SSA) to be carried out for targeting the

© Springer Nature Switzerland AG 2021

M. Mazzoleni et al., *Electro-Mechanical Actuators for the More Electric Aircraft*,
Advances in Industrial Control, https://doi.org/10.1007/978-3-030-61799-8_5

airworthiness certification is presented, with particular focus on Functional Hazard Assessment, Fault-Tree Analysis, Failure Mode Effects, and Criticality Analysis. A practical example of Preliminary SSA (PSSA) on the electro-mechanical actuation system for the morphing flaps of a more electric aircraft demonstrator is given.

Chapter 3 reviewed the basic concepts and terminology of fault diagnosis and condition monitoring approaches. The most common methodologies are exemplified, mostly focusing on model-based, signal-based, and knowledge-based approaches. These methods will be further investigated in Chap. 4 in the context of various experimental projects. The most common flight actuator faults are described with their specific terminology and results for flight EMAs from scientific literature are collected.

Chapter 4 is the main contribution of the monograph: it presents four research projects investigating the development of fault diagnosis, condition monitoring, fault prognosis, and fault-tolerant control of flight EMAs.

The first project is HOLMES (Sect. 4.2.1): it focuses on fault detection algorithms for secondary flight control surfaces actuators. After a fault injection step, that consisted of removing materials from the balls inside the nut of a ball screw transmission, a knowledge-based method based on machine learning classifiers is presented and evaluated on experimental data (Sect. 4.5.1).

Another relevant R&D activity is described with reference to the REPRISE project, which can be divided into two phases. In the first phase (Sect. 4.2.2), a condition monitoring activity has been carried out on a 1:1 scale EMA for primary flight control surfaces. A large experimental activity was carried out in order to progressively bring the actuator (with special focus on the ball screw transmission), to failure. In order to accelerate the degradation process, it was necessary to reduce the lubrication of the ball screw nut, while, at the same time, running the actuator with excessive load with respect to the nominal tolerated one. Two knowledge-based approaches were presented. The first one is based on change detection methods (Sect. 4.5.2), while the second one relies on statistical process control techniques (Sect. 4.5.3).

In the second phase of the REPRISE project (Sect. 4.2.3) a fault diagnosis and fault-tolerant control scheme applied to an innovative fault-tolerant flight EMA is presented, using high-fidelity dynamic models (Sect. 4.3.3) for jamming FDI and EMA condition monitoring.

The last project is related to primary flight control EMAs for medium altitude long endurance unmanned aerial vehicles (Sect. 4.2.4). In this context, two model-based approaches are presented: one based on reduced dynamic models for fault diagnosis and condition monitoring (Sect. 4.3.1) and one based on high-fidelity dynamic models for fault prognosis (Sect. 4.3.2). The first approach was tested by simulating electrical faults, while the second one was tested by simulating different levels of freeplay for mechanical fault prognosis.

Signal-based approaches are presented in Sect. 4.4 in a more general way, since only a small portion of existing projects on flight EMA, in the literature, have been faced relying on these techniques. First, the types of fault diagnosable using signal-based approaches are reviewed. These are mostly devoted to bearings and electrical faults, mainly approachable by looking for specific frequencies in the spectrum of some known signal (i.e., vibration or current measurements). A practical example, regarding the methodological flowchart for detecting and isolating localized bearings faults, is presented in Sect. 4.4.2.

5.2 Lessons Learned: Notes for Practitioners

The experimental projects presented in this book faced several preparation steps before digging into the experimental phase. Even during acquisition trials, it was necessary to constantly check the progression of the system situation. This section aims to collect lessons learned and experience gathered during the execution of the diagnosis projects faced by the authors.

5.2.1 Problem Definition

Generally speaking, a diagnosis problem formulation involves three main steps:

1. definition of the aims and objectives of the method that has to be developed;
2. data gathering;
3. development of the algorithms.

The value of these words extends above the specific application to a domain, but could be of general interest to any (mainly electro-mechanical) application.

STEP 1: Definition of the aims and objectives. It is mandatory to have a clear idea of what the algorithms have to accomplish, what are the most critical components to be monitored, and if all (presumed) required information are available. Summarizing, one may ask:

- What answer do I want to get from the algorithm?
  - *Fault diagnosis*: fault / no-fault decisions, location and entity of the faults;
  - *Condition monitoring*: monotonic indicators of machine health status;
  - *Fault prognosis*: monotonic indicators of machine health status and forecast of future health states.
- Do I have all the information needed?
  - If present, collect a history of past faults occurrences (maintenance activity reports are a very important information to understand the specific criticalities);
  - evaluate what physical quantities are useful to characterize the system operational behaviors, both in healthy and faulty states;

- What are the most critical components?
  - Perform FMECA, FTA or qualitatively evaluate the criticality and occurrence frequency of faults on the system components;
  - exchange information with the maintenance department to collect important information about the history of maintenance problems and reports.

STEP 2: Data gathering. Data is an essential element for the development of a diagnosis or monitoring algorithm. Data are very important independently of the specific method you will choose. They are obviously more significant if knowledge-based or signal-based approaches will be used, but they are necessary also in model-based approaches in order to estimate unknown parameters (system identification) and to align the model with the real process.

For this reason, it is important to think beforehand about which data (and how) to acquire. Then, one can ask the following questions:

- How to define the test plan?
  - Definition of the typical operation of the machine or its working regimes;
  - execution of tests in healthy state and tests in faulty state;
  - definition of what represents a faulty state: it is necessary to artificially introduce faults (fault injection) or an endurance session is needed?
- How to acquire the measurements?
  - Think about the need for additional sensors. If so, they must be acquired with specific hardware that has to be chosen or bought;
  - there may be the necessity to synchronize data from multiple different data sources;
  - the amount of data to be stored can be relevant: think about the need to expand memory storage;
  - if possible, check at the end of the day (even during) what has been acquired and if it is in line with expectations.

STEP 3: Development of the solution. Once data are acquired and stored, one has to select the most convenient techniques for the data at disposal.

- What approach is best suited to solve my problem with the acquired data?
  - **Model-based approach**: based on a mathematical model of the machine or component. Here, it is important to carefully evaluate the available time and cost to develop the modeling and system identification activities;
  - **Signal-based approach**: based on a symptom of the fault which is visible within a specific signal;
  - **Knowledge-based approach**: based only on the observed data (statistical methods). It is supposed that there are symptoms of faults in the data, but they need to be discovered, by

  · *Supervised methods*: estimate the relation from data to health condition;
  · *Unsupervised methods*: employ anomaly detection methods such as the Local Outlier Factor (LOF) method [2].

  Knowledge-based methods require to perform an experimental campaign for data collection in different regimes.

- How to deploy the solution?
  - If possible, the technological limitations of the computing hardware should be known beforehand. The software development team has to be included in the definition of the fault diagnosis algorithm, in order to share information about the hardware resources that will be available.

5.2.2 *Practical Considerations*

Diagnosis and monitoring applications. Analytical fault diagnosis and condition monitoring are foreseen to be used widely in the upcoming years [4], since

- theoretical foundations for model-based methods are well established, signal-based approaches are effective when the fault symptoms are known and knowledge-based techniques can leverage the progress trend in data science and machine learning;
- usually, fault diagnosis and monitoring algorithms do not influence the stability of the controlled system. Thus, they do not interfere with existing control software certifications;
- innovative technological solutions may bring additional complexity to the system, therefore, requiring more advanced diagnosis schemes.

A fault diagnosis method should be easily tunable to different situations even by non-experts (for example by trading-off alarm rates and missed detections). Furthermore, diagnosis and monitoring methods should be interpreted as a way to enhance the pilot and crew situation awareness, which is an important aspect inmanaging critical situations. Usually, a Human-Machine Interface (HMI) is responsible for presenting the whole system status to the operators by binary (OK/NOT OK) information (also continuous information, such as the fuel level, can be displayed). Condition monitoring approaches can alert the crew of possible subsystem problems that are developing during the flight, with the aim to allow the crew to react more timely to the situation.

The myth of the faulty condition. The evaluation of a diagnostic algorithm requires to test it against both a "normal" (healthy) and a "deviation from normal" (faulty) condition. While the nominal condition is easy to obtain (which encompasses obvious tolerances on the manufacturing of the system), the faulty condition on EMAs is a much more difficult state to conceive and obtain. First of all, the scarcity of experiments and employment on flight EMAs does not contribute to build knowledge about which types of faults are more probable than others. Secondly, even when there is a somewhat clear idea of what a faulty condition means, its replication is far from easy to reproduce artificially, either injected or induced through endurance tests.

As an example, consider developing a condition monitoring algorithm for flight EMAs. In this case, the most critical components selected for degradation have to be selected. Suppose that the mechanical transmission is of interest. Given the very high requirements that the final product have to meet, it is virtually impossible to degrade the transmission (e.g., ball screw and nut assembly) in times compatible with the project. For this reason, some interventions may be foreseen, such as:

- accelerate the degradation process by reducing the lubricant level: most of the times this is also a plausible situation (e.g., due to a maintenance error or a natural lubricant dwindling);
- run the system with excessive loads: in this case, a Finite Element Method (FEM) analysis should be conducted on the transmission in order to study the effect of those load forces on its components.

Despite these and other smart solutions, it is sometimes very difficult to study the degradation phenomenon within its natural time evolution.

Fault injection is another quite often employed solution to study a faulty condition. Then, one should ask if the injected fault:

- is meaningful or plausible;
- can be measured as regards its "intensity level" or "harshness".

Summarizing, testing for a faulty condition is a much more difficult step than sometimes thought at first instance.

Availability of a dedicated test bench. While in other industrial sectors it is sometimes possible to drive tests on the final system itself, in aerospace industry experimental tests, especially those involving injections of faults or degradation of system components, are not possible to perform if a test bench is not present.

Some care should be paid in the employed test bench setup. If the rig is reused from a previous (and maybe quite different) project, one has to be certain that proper modifications are put in action to adapt the rig to the new project. This intervention has to be seen as a strategical investment: if the rig presents problems (e.g., low quality measures, measurement and command software that is error-prone) the whole project can go wrong.

If it is possible, the types of tests that have to be performed should be carefully defined before finalizing the setup of the test bench, in order to modify the control and the human interface software to accept as input the desired tests.

A great simplification of the work is made possible if the test bench and the measured data are accessible from distance (online access), by the person who has to perform the tests and gather the data, in order to reduce the overheads and time waste. It is, however, necessary to have a reference person who can take care of connection or bench issues, to avoid too much wasted time.

If a test bench is not available for testing the faulty states of the system, high-fidelity models, capable of *simulating* faulty conditions, can be leveraged as advocated in Sects. 4.3.1 and 4.3.3. This means that faulty data will be simulated by a mathematical (software) model.

Furthermore, the test bench should be designed to be more robust than the device under test, such that the endurance session will not break the bench before than the device of interest.

The test bench should be easily modifiable, since some requirements could emerge during the work and during the data analysis.

Availability of measurements. It is useful to log, and thus have at disposal from the acquisition setup, all the variables that the ECU of the EMA has available during its normal operating mode. Also, measurements that will not be available in normal operating conditions, cannot be used as inputs to the diagnosis or monitoring algorithm.

During endurance campaigns, it is sometimes impossible to store (and then analyze) all the continuously-measured data, because it would require to store a large number of variables logged with high frequency for a long period of time. Thus, it has to be decided *when* and *for how long* to acquire the data (monitoring trials) and when not to acquire the data but *only degrade* the system (degradation trials). If an imminent failure of the system is envisaged (from a known number of operations performed or from having analyzed the acquired data), it can be desirable to have more frequent acquisitions, in order to better characterize the final degradation stage (which would probably be where most of the changes in the data can be observed, in a quite low amount of time).

An envisaged approach to fault diagnosis of airborne EMAs. Following the presented solutions to the fault diagnosis projects in Chap. 4, the following combined strategy can be suggested for the diagnosis of airborne EMAs:

1. the use of a **model-based approach** for *online* diagnosis and monitoring as presented in Section 4.3. Model-based approaches are able to decouple or attenuate the effects of noise and disturbances on the generated fault residuals. This is important due to the varying loads that affect the actuator in its different positions maneuvers and extensions. This enhanced robustness of the method against the variations of external conditions makes it ideal for (continuous) online monitoring of the actuator.
2. the use of a **knowledge-based approach** for *pre-flight* tests, as proposed in Sect. 4.5. These methods require that the external conditions do not change (or change only a little) with respect to those on which they are trained upon. For these reasons, they can be applicable when the aircraft actuator is not subject to external disturbances such as wind gusts. In this context, a pre-flight test can be interpreted as an IBIT test (or, if more time is required, as a MBIT test), see Sect. 2.3.5.

Signal-based approaches are actually applicable to a subset of all the possible actuator faults, specifically the ones for which a fault symptom is known. While they can possibly be robust against external disturbances such as load variations, their use in EMAs is limited to few specific components (i.e., bearings or some electrical faults.)

5.3 Other Possible Fault Diagnosis Activities for Airborne EMAs

This book is mainly focused on the fault diagnosis of flight EMAs, especially in its mechanical components. It was obviously not possible to discuss all possible failure cases in this context. Other works, such as [4] has a very similar structure like this book: it reports several projects related to various aspects of aerospace vehicles, two of which are related to flight EMAs. The interested reader can thus find it as a useful addition to explore even more possibilities and ideas.

One of the key points raised by the authors, to motivate the employment of analytical diagnostic techniques, is related to the possibility to early detect and isolate a fault from its originating envelope or component. This is because, if not detected and "stopped" with some strategy, the fault can propagate its effect through the structure of the aircraft. This "additional load" requires a reinforced structure to be tolerated. If the propagation can be stopped, the design of the structure can be lighter, thus saving weight and helping to attain sustainability goals (fuel burn, noise, range, and environmental footprint).

One of the presented projects is related to the diagnosis of an Oscillatory Fault Case (OFC) failure. OFC is an abnormal oscillation of a control surface due to component malfunction in control surface servo-loops. The vibration can propagate and excite the system structure. OFCs are considered as harmonic signals with frequency and amplitude uniformly distributed generally over the frequency range 0.1–10 Hz. Beyond 10 Hz, OFCs have no significant effects because of the low-pass behavior of the actuator. The state of practice method, implemented in Airbus A380 aircraft, consists of comparing the actuator position with the position simulated by a nonlinear model [3]. A more advanced approach has been proposed in [1], which makes use of a nonlinear observer for OFC fault detection and a sliding-mode approach for fault estimation.

The second project presented in [4] deals with the diagnosis of elevator runaway and jamming. Low-speed runaway degrades the aircraft's controllability and increases the pilot's workload. High-speed runaways are less problematic from a trajectory point of view, but lead to additional loads that must be taken into account in the aircraft structural design objectives. The jamming of an aircraft control surface creates a dissymmetry in the aircraft configuration, which must be compensated by appropriate deflections of other control surfaces. State of practice methods for detecting those faults consist of inconsistency checks between two or more available signals. For the runaway and jamming faults, the authors propose the use of a Kalman filter, where the employed model describes the time evolution of the runaway fault (e.g., like a sensor drift).

5.4 Future Perspectives

Academic research on fault diagnosis continues to offer new findings and methods, and it is fundamental to foster collaboration and research projects with industrial partners, in order to close the currently existing gap between research and practical implementation of techniques and methods on airworthy systems and equipments. A relevant difficulty is represented by the high robustness required to a monitoring system for aerospace applications (e.g., the system must be unsusceptible to false alarms or dormant failures when external conditions change, such as loads or temperature).

A possible approach for enhancing the monitoring systems robustness, with special reference to the minimization of dormant failures, could be the increase of detail of monitoring checks before the mission (i.e., during pre-flight built-in tests), when the uncertainties on external conditions are more limited.

In addition, aiming to minimize false alarms in flight (i.e., via the continuous built-in-tests), a combination of monitoring techniques can provide the optimal approach. Model-based real-time monitoring is probably the best solution for supervising the overall effectiveness of safety-critical equipments, but it can be not adequate for fault isolation. In this perspective, signal-based and knowledge-based techniques are expected to preferable, provided that the available computing resources in the equipments' electronic units are sufficient.

References

1. Alcorta-Garcia E, Zolghadri A, Goupil P (2011) A nonlinear observer-based strategy for aircraft oscillatory failure detection: A380 case study. IEEE Trans Aerosp Electron Syst 47(4):2792–2806. https://doi.org/10.1109/TAES.2011.6034665
2. Breunig MM, Kriegel HP, Ng RT, Sander J (2000) Lof: identifying density-based local outliers. In: Proceedings of the 2000 ACM SIGMOD international conference on Management of data, pp 93–104. https://doi.org/10.1145/335191.335388
3. Goupil P (2010) Oscillatory failure case detection in the a380 electrical flight control system by analytical redundancy. Control Eng Pract 18(9):1110–1119 https://doi.org/0.1016/j.conengprac.2009.04.003
4. Zolghadri A, Henry D, Cieslak J, Efimov D, Goupil P (2014) Fault diagnosis and fault-tolerant control and guidance for aerospace vehicles. Springer

Glossary

The terminology used in the diagnosis and fault-tolerant control literature has only during the recent years approached a coherency in the published material. In this book, we adhere to the terminology used in the current publications in the control systems community, see Chap. 3.

Active fault-tolerant system A fault-tolerant control system where faults are explicitly detected and accommodated. Opposite to a passive fault-tolerant system.

Analytical redundancy Use of two or more, but not necessarily identical ways to determine a variable, where one way uses a mathematical process model in analytical form.

Availability Likelihood that a system or an equipment will operate satisfactorily and effectively at any given point in time. $A = \frac{\text{MTTF}}{\text{MTTF} + \text{MTTR}}$, MTTR = Mean Time To Repair = $1/\mu$, μ is the rate of repair.

Burn-in faults Faults related to design errors or materials' imperfections. They occur in the initial phases of components' life.

Common Cause Analysis Analysis performed throughout the safety processes to identify potential common-mode faults or single-point-of-failures, via Zonal Safety Analysis, Particular Risk Analysis and Common Mode Analysis.

Condition monitoring A continuous task of determining the conditions of a physical system, by recording information, recognizing and indicate anomalies in the system behavior. The output of a condition monitoring algorithm are continuous indicators of the degree of the system health state, contrary to fault detection that produces a dichotomous output.

Data voting The process of obtaining a unique consolidated value of a data from multiple measurements or estimations of it.

Dependability Combination of reliability, availability and safety. It may also include recoverability, maintainability, maintenance support performance, durability and security. A dependable system is a fail-safe system with high availability and reliability.

© Springer Nature Switzerland AG 2021

M. Mazzoleni et al., *Electro-Mechanical Actuators for the More Electric Aircraft*, Advances in Industrial Control, https://doi.org/10.1007/978-3-030-61799-8

Digital twin A mathematical model of the system considered for fault diagnosis and monitoring. It may encompass also entire production lines, which behavior can be simulated for optimization purposes.

Disconnection fault The actuator connection with the load is not obtained as designed. Also known as free-play or free-floating.

Discrepancy An abnormal behavior of a physical value or inconsistency between more physical values and the relationship between them.

Durability In database systems, durability is the property which guarantees that transactions that have committed will survive permanently, even if the system crashes.

Error Deviation between a measured or computed value (of an output variable) and the true, specified or theoretically correct value.

Fail-safe A system having the capability to respond to a failure by reverting to a safe passivation/shutdown with no, or minimal, harm to other equipments, environment or persons.

Fail-operational The ability to sustain any single point failure.

Failure Permanent interruption of a system/component ability to perform a required function under specified operating conditions.

Failure effect The consequence of a failure mode on the operation, function, or status of an item.

Failure mode Particular way in which a failure can occur.

Failure Mode and Effect Analysis A table where the failure modes of each system part are classified and qualitatively analyzed in terms of effects to higher-level an lower-level parts.

Failure Mode, Effects, and Criticality Analysis A table where the failure modes of each system part are both qualitatively analyzed in terms of effects to higher-level an lower-level parts and quantitatively evaluated in terms of probability of occurrence.

Failure Mode and Effect Summary A table where the major FMECA results are reported, by highlighting the most relevant outcomes to be addressed for airworthiness certification requirements.

Failure probability Probability that, at time t, the fault is occurred.

Fault Unpermitted deviation of at least one characteristic property or parameter of a system from its acceptable/usual/standard condition. A fault is the occurrence of a failure mode. Failures and malfunctions originate from a fault.

Fault accommodation The action of changing the control law in response to faults, without switching off any system component. In fault accommodation, faulty components are still kept in operation thanks to an adapted control law.

Fault compensation The process of actively intervening to modify the system configuration after a fault, aiming to recover some level of system performance. Contrary to Fault accomodation, where no component is turned off, here the control law is changed and also the faulty component is deactivated.

Fault detection Determination if there are faults present in a system and time of detection.

Fault detector An algorithm that performs fault detection and isolation.

Fault diagnosis Determination of kind, size, location, time of occurrence of a fault and the fault signal. Fault diagnosis includes fault detection, isolation, estimation, and identification.

Fault estimation Reconstruct the time-varying behavior of the fault signal. Follows fault isolation.

Fault identification Determination of the size and time-varying behavior of a fault. Follows fault isolation.

Fault isolation Determination of the location of a fault, i.e., of the component that is faulty. Follows fault detection.

Fault prognosis Determination of the system future health state, given the current estimate. It usually build upon condition monitoring. Thus, its nature is iterative, and the prediction gets updated each time stamp as new data are available.

Fault recovery The result of a successful fault accommodation or system reconfiguration.

Fault-Tree Analysis Logical flowchart, which defines the dependence between the faults to system parts and a specific failure case.

Fault-tolerant system A system where a fault is recovered with or without performance degradation, but a single fault does not develop into a failure on subsystem or system level.

Feature Characteristic, attribute of a system that is influenced proportionally by the size (entity) of a fault. It is a manifestation of the presence of a fault. It can be measured and compared to its values in normal operation.

Free-play fault see Disconnection fault.

Free-floating fault see Disconnection fault.

Functional Hazard Assessment Qualitative analysis aiming to define the potential hazards related to the loss of functional requirements in specific mission phases.

Hard-over fault see Runaway fault.

Hardware redundancy Use of more than one independent physical instrument to accomplish a given function.

Hazard rate The increment of fault probability referred to the components survived at time t.

Incipient fault A fault where the effect develops slowly, e.g., clogging of a valve. In opposite to an abrupt fault.

Jamming fault The actuator is stuck in a position and can no longer move.

Lock-in-place fault see Jamming fault.

Loss of effectiveness fault The actuator does not track well the control demand with adequate performances.

Maintainability The ease with which a product or system can be maintained.

Maintenance support performance The ease with which a product or system can be maintained.

Malfunction A malfunction is an intermittent irregularity in the fulfillment of a system desired function. The only difference with a failure is that, in the malfunction case, the interruption of the system function is only temporary. The (stable) degradation of the system performance can be considered as a malfunction.

Monitor Algorithm that performs diagnosis or monitoring actions.

Monitoring see Condition monitoring.

More Electric Aircraft The gradual replacement of on-board systems based on mechanical or pneumatic power sources with electrically-powered systems.

More Electric Propulsion The gradual replacement of propulsion systems with electrically-powered ones.

Oscillatory Failure Case fault The actuator output deviates from control demand by exhibiting abnormal oscillations.

Perturbation An input acting on a system which results in a temporary departure from stady state.

Preliminary System Safety Assessment Analysis that supports the system architecture design, in which diverse systems' architectures, where diverse systems' architectures are compared in terms of RAMS features.

Quantitative model A system model describing the behavior with relations among system variables and parameters in analytical terms such as differential or difference equations.

Random faults Faults related to non-deterministic factors (e.g., overloads). They occur throughout the components' life.

Real-time model-based monitoring Use of models executable in real-time by the EMA ECU, in order to permit the model outputs to be synchronized and available together with the ones derived from hardware components.

Reconfigurability The possibility to recover a fault by using the reconfiguration strategy: switching off the faulty components, and changing the control law so as to achieve the specified objective by using only the healthy components.

Recoverability The property of system to being able to recover or being recovered.

Reliability Probability of a system to perform a required function under stated conditions, within a given scope and during a given period of time. Measure: MTTF = Mean Time To Failure. MTTF = $1/\lambda$; λ is the constant rate of failure [e.g., failures per hours].

Reliability Block Diagram A logical diagram which defines the dependence between the reliability of system parts and the reliability of a specific system function.

Remedial action A correcting action (reconfiguration or a change in the operation of a system) that prevents a certain fault to propagate into undesired end-effects.

Residual Signal that carries fault information, based on deviation between measurements and model-based computations.

Residual generator Typical of model-based fault diagnosis, it is the component (often a dynamic system) which produces residuals based on measured values of the inputs and outputs of the system.

Runaway fault The actuator moves without control demand towards its endstroke. Also known as hard-over fault.

Safety Ability of a system not to cause danger to persons or equipment or the environment.

Securability The characteristic or degree of being securable, especially the ability of a system to provide different levels of secure access.

Severity A measure on the seriousness of fault effects using verbal characterization. Severity considers the worst-case damage to equipment, damage to environment, or degradation of a system's operation.

System Safety Assessment Analysis that supports the system design and implementation, in which the developed hardware and software are analyzed/verified via FTA and FMECA.

Supervision Monitoring of a physical system and taking appropriate actions to maintain the operation in the case of faults.

Survivability probability The probability that, at time t, the fault is not occurred.

Supervisor Algorithm that performs supervision.

Symptom A change of an observable quantity from normal behavior. Signal-based diangostic approaches look for specific fault symptoms in predetermined signals.

Stall fault The actuator dynamics is characterized by cyclic to intermittent saturation phenomena.

Threshold Limit value of a residual deviation from zero, so if exceeded, a fault is declared as detected.

Wear-out faults Faults related to materials' aging. They occur in the final phases of components' life.